"GOVERNMENT FROM REFLECTION AND CHOICE"

Constitutional Essays on War, Foreign Relations, and Federalism

CHARLES A. LOFGREN

New York | Oxford
OXFORD UNIVERSITY PRESS
1986

Oxford University Press
Oxford New York Toronto
Delhi Bombay Calcutta Madras Karachi
Petaling Jaya Singapore Hong Kong Tokyo
Nairobi Dar es Salaam Cape Town
Melbourne Auckland

and associated companies in
Beirut Berlin Ibadan Nicosia

Published by Oxford University Press, Inc.,
200 Madison Avenue, New York, New York 10016

Library of Congress Cataloging-in-Publication Data
Lofgren, Charles A.
"Government from reflection and choice."
Includes index.
1. War and emergency powers—United States.
2. United States—Foreign relations—Law and legislation.
3. Executive power—United States.
4. United States—Constitutional law.
I. Title.
KF5060.L64 1986 342.73'04 86-2357
ISBN 0-19-504007-4 347.3024

Printing (last digit): 9 8 7 6 5 4 3 2 1

Printed in the United States of America
on acid-free paper

"GOVERNMENT FROM REFLECTION AND CHOICE"

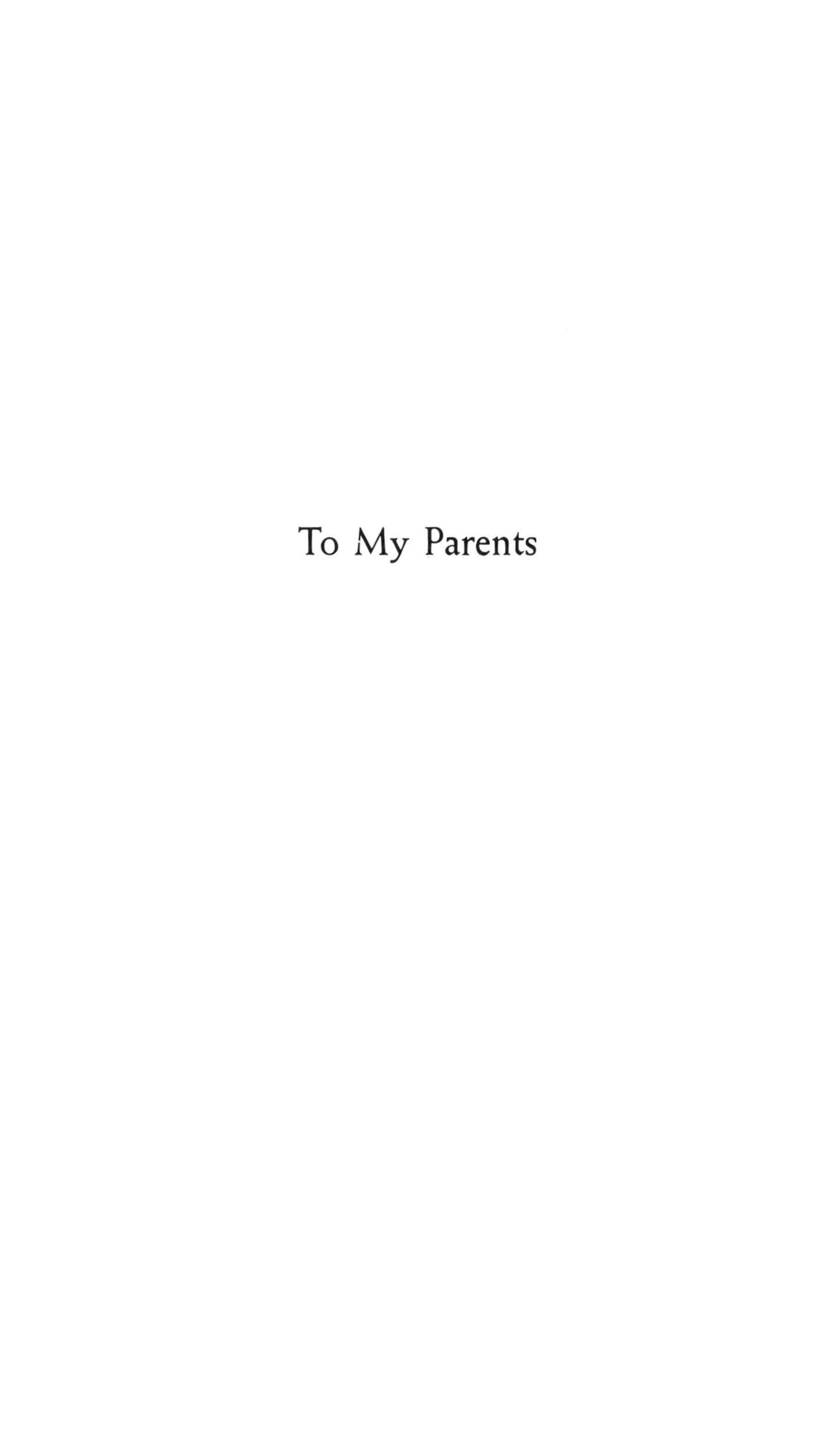

To My Parents

ACKNOWLEDGMENTS

Grateful acknowledgment is made of permission to reprint these essays:

"War-Making Under the Constitution: The Original Understanding," reprinted by permission of The Yale Law Journal Company and Fred B. Rothman & Company from *The Yale Law Journal,* Vol. 81, pp. 672-702.

"Compulsory Military Service Under the Constitution: The Original Understanding," *The William and Mary Quarterly,* Third Series, Vol. 33 (1976), pp. 61-88; copyright assigned by the publisher to the author.

"The Origins of the Tenth Amendment: History, Sovereignty, and the Problem of Constitutional Intention," reprinted by permission of Southwestern University School of Law from *Constitutional Government in America: Essays and Proceedings from Southwestern University Law Review's First West Coast Conference on Constitutional Law,* Ronald K. L. Collins, editor (Carolina Academic Press, 1980), pp. 331-357. Copyright © 1980 by Southwestern University Law School.

"Missouri v. Holland in Historical Perspective," reprinted by permission from *The Supreme Court Review,* 1975, pp. 77-122, edited by Philip B. Kurland, published by the University of Chicago Press, copyright © 1976 by The University of Chicago.

"*United States v. Curtiss-Wright Export Corporation:* An Historical Reassessment," reprinted by permission of The Yale Law Journal Company and Fred B. Rothman & Company from *The Yale Law Journal,* Vol. 83, pp. 1-32.

"Mr. Truman's War: A Debate and Its Aftermath," reprinted by permission from *The Review of Politics,* Vol. 31 (1969), pp. 223-241, published by the University of Notre Dame Press, Notre Dame, Indiana.

CONTENTS

INTRODUCTION

American constitutional history has a subject that doubles as its special and long-lived client: the Constitution itself. While this observation helps account for the durability of the field, it risks offending those practitioners who describe their work differently; but I ask them to bear with me for the moment. As a more serious consequence, it may tempt historians of all sorts to respond with a two-count indictment. For one thing, they may argue, the formulation hints at an anachronistic departure from simply trying to discover the truth about the past, which is still not a bad "first cut" at defining the historian's task. Second, by conjuring up a perhaps non-existent and certainly non-human entity called "the Constitution," which becomes an historical actor, it also suggests the fallacy of reification. What I have in mind, however, is something different on both counts. An explanation provides a preliminary view of the essays in this volume.

Consider the proposition that the Constitution is a client. Far from compromising the historian's integrity, this characteristic of the field reflects a concern for accurate history. Despite all the talk among constitutional scholars–and among "academic lawyers" in particular–about the need to break away from interpreting the Constitution itself in order to explain and apply the Constitution, judges and others often (and sensibly) ignore the advice.

Willy-nilly, it seems, history figures into constitutional debates, even though, as Justice Robert Jackson observed when assessing President Harry S Truman's seizure of the steel mills, "Just what our forefathers did envision, or would have envisioned had they foreseen modern conditions, must be divined from materials almost as enigmatic as the dreams Joseph was called upon to interpret for Pharaoh."[1] Although the sources are not always as enigmatic as Jackson indicated, this penchant for historical argument creates a still-pressing need for careful work in American constitutional history, and no less today than a quarter-century ago when Paul Murphy challenged historians to reclaim the field.[2] Even the "noninterpretivists" or "nonoriginalists," who attack reliance on original meanings and original intentions in the course of deciding real cases in the present, advance a position more sophisticated than the one caricatured at the beginning of this paragraph; for they, too, seem to recognize the usefulness (if not the conclusiveness) of historical inquiry.[3]

Ultimately, however, the fact that constitutional history may have a policy dimension is only a bonus. Once discerned, the issues are primarily interesting because they are there, as I explain toward the beginning of Chapter Three; and what issues! They emerge from problems that turned out to engage some of the best minds in our history and had outcomes affecting the experiences of practically everyone, at times in petty detail, yet often in ways fundamental to human integrity, and on occasion with tens or hundreds of thousands of actual lives (and lately, more) hanging in the balance. Not least, they frequently stimulated reasoned debate over the true purposes and ongoing direction of the American experiment.

This leads us to the second count in the hypothetical indictment. Is the focus of constitutional history a nonexistent abstraction? I think not. True, the Constitution framed in 1787, ratified

1. Youngstown Sheet and Tube Company v. Sawyer, 343 U.S. 579, 634 (1952).

2. Paul L. Murphy, "A Time to Reclaim: The Current Challenge of American Constitutional History," 69 *American Historical Review* (1963), 64-79.

3. See Paul Brest, "The Misconceived Quest for the Original Understanding," 60 *Boston University Law Review* (1980), 204-38.

in 1787-88, subsequently amended some twenty-six times (as of 1985), and explicated in thousands of volumes of court reports, may correctly be described as an *artificial* entity, but in an original sense of the term–"produced by man." Moreover, the historian's constitution (and here I intentionally use a lower-case "c") embraces more than the 1787 document, its amendments, and judicial commentary about it. Lord Bolingbroke's description, in 1733, of the English constitution provides a better guide:

> By constitution we mean, whenever we speak with propriety and exactness, that assemblage of laws, institutions and customs, derived from certain fixed principles of reason, directed to certain fixed objects of public good, that compose the whole system, according to which the community hath agreed to be governed.[4]

As a definition, this statement admittedly fails to offer a simple criterion for distinguishing the "constitutional" realm from the more ephemeral; and questions about boundaries may be tough to resolve. (Such questions also fail to absorb much of the constitutional historian's energy–and rightly so, as evidenced lately in the cross-disciplinary enrichment provided by the neighboring field of legal history.[5]) But Bolingbroke's view has the virtue of calling our attention to the central role of men and women, what they thought and what they did, what they expected and what they ruled out (and what they overlooked), how they related together and organized themselves politically and understood what they were deciding and doing. In short, it reminds us that we confront a human institution, with all the resulting messy complexities. It also reminds us, or should, that understanding the "is" of the past may require the constitutional historian to probe seriously the importance given the "ought"–those "fixed principles of reason" and "fixed objects of public good"–by the participants in the constitutional process.

4. Quoted in Charles H. McIlwain, *Constitutionalism, Ancient and Modern* (rev. ed., 1947; Ithaca, N.Y., 1966), 3.

5. See, for example, Harry N. Scheiber, "American Constitutional History and the New Legal History: Complementary Themes in Two Modes," 68 *Journal of American History* (1981), 337-50.

At minimum, the constitutional historian strives to keep in mind the easily forgotten truism that a document itself can intend nothing. It is unquestionable, of course, that words and phrases—"to declare war," "armies," "expressly," "reserved to the States," "police action," to mention a few that enter into these essays—sometimes take on singular importance. But the interesting question is generally their nuanced and shifting meanings to actual people in the past. Undoubtedly, too, specific documents that hold (or held) authoritative legal status—whether the United States Constitution or a host of others—offer clues to what people thought, desired, and intended, but rarely do they provide all the necessary evidence.

Indeed, constitutional historians share the frustrations experienced by historians of any stripe as they pose questions about (and to) the past. Regarding the most interesting problems, it seems, one never finds *all* the evidence necessary for constructing answers that will satisfy every careful critic. As Jack N. Rakove so nicely put it, after examining a topic related to several explored in the present collection, "One understands why Justice Jackson may have yearned for the inspiration of Joseph, but the historian interested in these issues finds something to be said for the patience of Job."[6]

So much for the indictment of the field in general. As for the essays in this collection, I confess that for several the catalyst was not an initial puzzlement about the more distant past, but rather the debates over the war in Vietnam, which was the "present" when they were written or at least conceived. Accordingly, they may reveal my own "hawkish" attitudes during that lamentable period, but my sense of the matter is that the historical puzzle-solving which was inherent in doing them—and which gave their topics the fascination that finally hooked me—led me to conclusions quite at odds with my policy commitments at the time. The reader may judge.

I think I am on less disputable ground in identifying another aspect of these essays. Puzzles enliven the recent past no less than the more distant past; and so, in its elusiveness, the line between

6. Jack N. Rakove, "Solving a Constitutional Puzzle: The Treatymaking Clause as a Case Study," 1 *Perspectives in American History* (new series, 1984), 281.

the recent past, the present, and indeed the future, becomes a seductive temptress. At various points in this collection, unquestionably, the reader will find that I fell prey to her wiles. As a result, and particularly because I have resisted the further temptation, occasioned by the reissuance of these essays, to re-do parts of them, it seems useful to conclude this brief introduction with two lines of comment–or better, of afterthought. One relates to Chapters One and Six, which focus on the original understanding of war-making and Harry Truman's war; the other involves Chapter Three, on the origins of the Tenth Amendment.

As described at the end of Chapter Six (which was written in mid-1968), the Southeast Asian or "Tonkin Gulf" Resolution did not prevent domestic debate over the Vietnam War. The failure occurred even though a good case can be made that when judged against the original understandings of war-making (laid out in Chapter One), the Resolution provided sufficient constitutional warrant for President Lyndon B. Johnson's course of action. In contrast, by winding down the war even while the actual peace settlement (such as it was) did not come until January 1973, President Richard M. Nixon gained at least some of the domestic acquiescence that had eluded Johnson. This sequence *may* reinforce one of Chapter Six's conclusions, that the frustration of limited warfare, rather than the presence or absence of constitutionally adequate congressional authorization, was the real rub for wartime critics. Yet I suspect that such an assessment understates the impact of the constitutional issue itself; for despite the slackening of American involvement, the drive continued for legislation to regulate presidential war-making. There may have been, as I wrote in 1968, a consensus during the Korean War "that the President could act, without Congress, to counter an immediate, dangerous threat to American security," and that view may have persisted. But I was surely wrong in further claiming, "Thus the real issue became (and remains): What constitutes such a threat?" *An* issue, yes, but hardly the only real issue. Subsequent events are instructive.

In October 1973 the War Powers Resolution became law over President Nixon's veto. By then, American military involvement in all of Indochina had been ended, partly under threat (at last) of a funding cutoff by Congress. If not as restrictive as some congressmen urged (they alleged it unconstitutionally gave the President prospective authority for limited use of troops abroad), the Resolution nonetheless evoked attacks from others who charged that it sought unconstitutionally to limit presidential prerogative in employing forces to protect the national interest. In his veto message, Nixon branded the measure a rigid codification contrary to the Founding Fathers' clear understanding of the need for flexibility.[7] The terms of the debate, along with the language that has accompanied subsequent hedging over compliance with the War Powers Resolution by Presidents Ford, Carter, and Reagan, indicate that the constitutional dimensions of war-making, including the understandings of the founding period, still engage serious minds.[8] But here, too, closer attention to origins might contribute to clearing out a thicket or two. The understandings explicated in Chapter One, I would argue, bolster the constitutional propriety of *both* prospective authorization *and* regulation by Congress of "imperfect war," as eighteenth-century Americans labeled it.

War and foreign relations today loom as perhaps the major sources of actual or potential dispute over the respective powers of Congress and the President. In constitutional terms, they present *structural* issues. Where does authority properly lie within the federal or central government? But another sort of structural issue–federalism, or the division of power between states and nation–used to arouse Americans and occasionally still does.[9] The Tenth Amendment provides what amounts to the totemic text of

7. *Congressional Record,* 93 Cong., 1st sess., 34,990 (October 25, 1973). Nixon also attacked the constitutionality of allowing Congress to terminate military operations by concurrent resolution. *Ibid.,* at 34,990-91. This feature of the War Powers Resolution is probably unconstitutional under the Supreme Court's subsequent decision in *Immigration and Naturalization Service v. Chadha,* 103 S.Ct. 2764 (1983), which rejected the use of one-House and concurrent resolutions as "legislative vetoes."

8. For a useful summary of the War Powers Resolution and its background and subsequent applications, see Louis Fisher, *Constitutional Conflicts between Congress and the President* (Princeton, N.J., 1985), 307-18.

9. To be sure, questions involving federalism may also raise foreign affairs issues, as Chapter Four demonstrates by exploring *Missouri v. Holland,* 252 U.S. 416 (1920).

this problem area, and Chapter Three, which received its final touches in 1978, explores the amendment's origins. Again, though, the elusive line between past, present, and future proved an irresistible temptress, for the chapter opens and closes by baldly addressing a controversy over federalism that flared anew in 1976 after forty years of quiescence and now appears unlikely soon to disappear. Hence my second line of afterthought:

In 1976, *National League of Cities v. Usery* breathed new life into the doctrine that the federal system places limits on the powers of the central government. Although the Commerce Clause of the Constitution gives Congress plenary power over interstate commerce, and although the Supreme Court in *Usery* conceded that the congressional enactment under challenge involved a subject (the wages of state workers) affecting such commerce, a 5-4 majority nevertheless held that the regulation exceeded Congress's authority. At the end of the Tenth Amendment essay, I suggested that this logic in Justice William H. Rehnquist's opinion for the Court was problematic. I am still troubled by the argument, but now see the difficulty as more serious for a constitutional literalist than for a critic trying to adhere to original understandings.

In particular, and not without irony, Rehnquist's position received needed amplification in *Garcia v. San Antonio Metropolitan Transit Authority* (1985), in which a new 5-4 majority flatly overruled the 1976 decision. Rejecting the *Usery* reasoning that federal wages-and-hours regulation unconstitutionally compromised the ability of states *as states* to perform certain essential functions, the *Garcia* majority contended that the states were better protected through the political process than by judicial application of limitations associated with the Tenth Amendment.[10] But the dissenters mauled such reasoning, noting that *elimination* of judicial protection hardly left the states *better* protected, particularly when the supposed political safeguards of federalism had largely atrophied over the past several decades. One of the minority justices, Sandra Day O'Connor, added:

10. Garcia v. San Antonio Metropolitan Transit Authority, 105 S.Ct. 1005 (1985).

> It is worth recalling the . . . passage in *McCulloch v. Maryland* [1819] . . . that lies at the source of the recent expansion of the commerce power. "Let the end be legitimate, let it be within the scope of the constitution," Chief Justice Marshall said, "and all means which are appropriate, which are plainly adapted to that end, which are not prohibited, but consist with the letter *and spirit* of the constitution, are constitutional." . . . The *spirit* of the Tenth Amendment, of course, is that the States will retain their integrity in a system in which the laws of the United States are nevertheless supreme. [O'Connor's emphasis.]

Justice Rehnquist expressed his confidence that this principle would "in time again command the support of a majority."[11] So another debate over origins seems likely to continue–along with normally intense interest in new Court appointments.

Finally, though, I make no real apology for occasionally straying in these essays from strictly historical concerns. At the beginning of *Federalist* 1, Alexander Hamilton observed:

> It has been frequently remarked, that it seems to have been reserved to the people of this country, by their conduct and example, to decide the important question, whether societies of men are really capable or not, of establishing good government from reflection and choice, or whether they are forever destined to depend, for their political constitutions, on accident and force.[12]

Today's arena may be broader, but those are still the stakes.

11. *Ibid.,* at 1021-38 (dissents by Justices Powell, Rehnquist, and O'Connor). The quotation from O'Connor is at 1036; the quotation from Rehnquist (which O'Connor specifically endorsed in her opinion) is at 1033. The dissenters drew considerable sustenance from Robert F. Nagel, "Federalism as a Fundamental Value: *National League of Cities* in Perspective," 1981 *Supreme Court Review,* 81-109.

12. *The Federalist,* No. 1, Jacob E. Cooke, ed. (Middletown, Conn., 1961), 3.

"GOVERNMENT FROM REFLECTION AND CHOICE"

1

WAR-MAKING UNDER THE CONSTITUTION:
The Original Understanding

The first major public debate over the division of war-making power between Congress and the President occurred in mid-1793 following President Washington's proclamation of American neutrality[1] in the war which had broken out between England and France at the beginning of that year. Defending Washington's action in a series of newspaper articles under the disarming pseudonymn of "Pacificus," Alexander Hamilton, a participant in the Constitutional Convention six years earlier, argued that since war-making was by nature an executive function, Congress could exercise only those aspects of it which the Constitution specifically grants the legislature. These grants, being exceptions to the general rule, must be narrowly construed.[2] James Madison, a principal framer of the Constitution and co-author with Hamilton and John Jay of *The Federalist Papers,* found the Constitution equally clear, but to the opposite effect. Writing as "Helvidius," Madison asserted that war-making was a legislative function and that any

1. Proclamation of April 22, 1793, 1 AMERICAN STATE PAPERS: FOREIGN RELATIONS 140 (W. Lawrie & M. Clark eds. 1833).

2. *See* 4 THE WORKS OF ALEXANDER HAMILTON 437-44 (H. Lodge ed. 1904). Specific exceptions to the President's war-making power, as listed by Hamilton, were "the right of the Legislature 'to declare war and grant letters of marque and reprisal.' " *Id.* at 439. *See* U.S. CONST. art. I, § 8.

exceptions in favor of the executive must be strictly interpreted.[3]

The debate thus opened has continued sporadically to the present day. Most recently, limited wars in Korea and Indochina have occasioned renewed interest in the question of which branch of the federal government is constitutionally empowered to commence war.[4] In the course of this debate, however, constitutional scholars have generally failed to examine thoroughly the problem of how Americans in 1787-88 understood the war-making clauses of the Constitution.[5] Such scholarly reticence[6] doubtless has some

3. *See* 6 The Writings of James Madison 138-88 *passim* (G. Hunt ed. 1906). Under Madison's interpretation, the key exception to the war-making power of Congress was the provision that the President was "Commander in Chief of the Army and Navy of the United States, and of the militia when called into the actual service of the United States." *Id.* at 148. *See* U.S. Const. art. II, § 2.

4. The literature on this topic is enormous. Useful introductions may be found in *Hearings on War Powers Legislation Before the Subcomm. on National Security Policy and Scientific Developments of the House Comm. on Foreign Affairs,* 92d Cong., 1st Sess. (1971); *Hearings on Congress, the President, and the War Powers Before the Subcomm. on National Security Policy and Scientific Developments of the House Comm. on Foreign Affairs,* 91st Cong., 2d Sess. (1970); M. Pusey, The Way We Go To War (1969); Note, *Congress, the President, and the Power to Commit Forces to Combat,* 81 Harv. L. Rev. 1771 (1968); E. Corwin, The President: Office and Powers 1787-1957, History and Analysis of Practice and Opinion (4th rev. ed. 1957); House Comm. on Foreign Affairs, Background Information on the Use of United States Armed Forces in Foreign Countries, H.R. Rep. No. 127, 82d Cong., 1st Sess. (1951). For earlier studies, see J. Rogers, World Policing and the Constitution: An Inquiry into the Powers of the President and Congress, Nine Wars and A Hundred Military Operations 1789-1945 (1945); C. Berdahl, War Powers of the Executive in the United States (1921).

5. The clauses of primary significance are art. I, § 8, empowering Congress "[t]o declare War [and] grant Letters of Marque and Reprisal . . . ," and art. II, § 2, making the President "Commander in Chief of the Army and Navy of the United States, and of the Militia of the several States, when called into the actual Service of the United States . . . ." Of lesser importance is the clause of art. I, § 10, providing that "[n]o State shall . . . engage in War, unless actually invaded, or in such imminent Danger as will not admit of delay." The clauses of art. I, § 8, dealing with the raising, support, government, and regulation of land and naval forces—that is, the war-supporting clauses—are outside the scope of the present study. For debate surrounding them, see Donahoe & Smelser, *The Congressional Power to Raise Armies: The Constitutional and Ratifying Conventions, 1787-1788,* 33 Rev. of Pol. 202 (1971).

6. Recent commentators have discussed the original understanding, but regardless of their positions on present-day issues, they have generally limited such

claim to prudence since Madison and Hamilton, who presumably knew something about the original intent, came to contradictory conclusions within a few years of the Constitutional Convention. But the fact that two of the framers were in disagreement is thin justification for according this significant issue inadequate treatment. Hence, by examining the debates and proceedings which accompanied the framing and ratification of the Constitution, and particularly by considering ideas prevalent among Americans of that day as they interpreted the clauses in question, I shall attempt in this article to throw light on two important questions: (1) What was the original understanding respecting the allocation between the President and Congress of the general power to commence war? And (2) was that power understood to include the commencement of *undeclared* war?

---

discussions to an examination of the one debate in the Federal Convention on changing the wording of the clause giving Congress the power "to make war" so that it conferred power "to declare war." For this debate, see 2 THE RECORDS OF THE FEDERAL CONVENTION 318-19 (M. Farrand ed., rev. ed. 1937) [hereinafter cited as FARRAND, RECORDS]. *See, e.g., 1970 Hearings on Congress, The President, and the War Powers, supra* note 4, at 207, 211 (statements of J. Stevenson, Legal Advisor, Dep't of State, and W. Rehnquist, Ass't Att'y Gen.); Bickel *et al., Indochina: The Constitutional Crisis,* in 116 CONG. REC. 7117, 7122-23 (daily ed. May 13, 1970); PUSEY, *supra* note 4, at 44-47 (treating some other points, but only vaguely); Note, *Congress, the President, and the Power to Commit Forces to Combat, supra* note 4, at 1773 & nn.14, 16; Wormuth, *The Vietnam War: The President versus the Constitution,* in 2 THE VIETNAM WAR AND INTERNATIONAL LAW 710, 713-17 (R. Falk ed. 1969); R. HULL & J. NOVOGROD, LAW AND VIETNAM 170 (1968). *But see* Reveley, *Presidential War Making: Constitutional Prerogative or Usurpation?,* 55 VA. L. REV. 1243, 1281-85 and accompanying notes (1969) (citing other evidence from the convention itself); R. Russell, The United States Congress and the Power to Use Military Force Abroad, April 15, 1967, at 1-65 (unpublished Ph.D. dissertation in the Fletcher School of Law and Diplomacy Library, Tufts University). While Russell neglects several points discussed in the present article, *see especially* pp. 13-16 & pp. 22-23 *infra,* and give less attention to certain other items, *see especially* pp. 16-22 *infra,* his conclusions are generally consistent with mine. Deserving special mention is *Hearings on War Powers Legislation Before the Senate Comm. on Foreign Relations,* 92d Cong., 1st Sess. at 7-121 *passim* (1972) (statements and testimony of Professors H. S. Commager, R. Morris, and A. Kelly, Mar. 8-9, 1971), which I was able to examine only after the present article was written. This provides an excellent, albeit relatively brief, discussion of the original understanding of 1787-88 in the course of a broader review of the history of war-making in its constitutional dimensions.

I. Problems of Allocating Power: From the Articles of Confederation to the Constitution

Under the Articles of Confederation, Congress exercised both legislative and executive powers. Consequently, the Articles, in dealing with the war-making power, needed only to provide that the "United States in Congress assembled, shall have the sole and exclusive right and power of determining on peace and war." (An exception provided that individual states might engage in war if they were actually invaded or were threatened with imminent Indian attack.)[7] Like the Constitution, however, the Articles today appear somewhat ambiguous on their face as to whether a war conducted by the United States necessarily had to be a "declared" war. They conferred the "right and power of *determining* on . . . war," but also referred to certain acts the states might perform only after a congressional "declaration of war."[8] Thus, while contemplating that at least some wars would be "declared" in form, the document did not explicitly resolve the question of whether the nation might engage in other sorts of war.[9]

Whatever the case with the Confederation, in 1787 the Philadelphia Convention drafted a Constitution which provided for a federal government with distinct branches, thereby necessitating some attention to the allocation of the war-making power of the government. Yet, while the new Constitution increased the already severe limitations which the Articles of Confederation had placed on state war-making,[10] similar explicitness did not mark the divi-

7. Arts. of Confed. arts. VI, IX. These provisions relating to war-making in the final Articles of Confederation were virtually unchanged from the first draft of the document in July 1776. *See* 5 Journals of the Continental Congress 1774-1789, at 549-50 (1906).

8. Arts. of Confed. arts. VI, IX (italics added).

9. Since the Articles did authorize the granting of letters of marque and reprisal, they tacitly suggest that other sorts of hostilities were contemplated. Much depends, however, on the types of situation to which letters of marque and reprisal were thought applicable, and on whether such situations were to be classified as war. The same ambiguity arises with respect to the Constitution and is discussed at pp. 27-32 *infra*.

10. Under the Articles, states could grant letters of marque and reprisal after a congressional declaration of war, subject to congressional regulation, while

sion of war-making power between Congress and the President. Nor have the records of the debates in the Federal Convention proved an adequate guide in resolving the ambiguities inherent in the "plain words" of the Constitution, principally because the question of the manner in which the nation should be committed to war was not one of the chief concerns of the delegates in Philadelphia. Criticism of the Confederation government for its inability to support federal objectives, both domestic and foreign, had not included the complaint that the Confederation was deficient in its ability to commit the nation to war.[11]

The main report of the one debate which explicitly considered allocation of the war-making power occupies little more than one page out of the 1,273 which contain the printed records of the Convention.[12] This debate occurred on August 17, 1787, while the Convention was considering the clause of the draft constitution reported by the Committee on Detail on August 6 which gave Congress the power "to make war."[13] Charles Pinckney of South Carolina opened the debate by arguing that the legislature as a whole was too cumbersome a body to exercise this power. He contended that it should be vested in the Senate, which was smaller and would be more knowledgeable in foreign affairs, and which by virtue of its treaty power[14] had the authority to make peace. Pierce Butler, from Pinckney's own state, carried the latter's argument a step further and called for vesting the power in the President, but his proposal received no recorded support. Madison and Elbridge Gerry of Massachusetts then "moved to insert *'declare'*; striking out *'make'* war; leaving the Executive the Power to repel sudden attacks."[15] This wording, from Madison's

---

under the Constitution they are absolutely forbidden to make such grants. *Compare* ARTS. OF CONFED. art. VI, *with* U.S. CONST. art. I, § 10.

11. *See* Farrand, *The Federal Convention and the Defects of the Confederation,* 2 AM. POL. SCI. REV. 532, 535-37 (1908).

12. *See* 2 FARRAND, RECORDS, *supra* note 6, at 318-19.

13. *Id.* at 181-82.

14. At this point in the Convention's proceedings, the Senate still had exclusive authority to make treaties, the President not yet having been given power to make treaties with the advice and consent of the Senate. *See id.* at 183, 392-94, 498, 538.

15. *Id.* at 318 (emphasis added).

notes, suggests that the change from "make" to "declare" was intended in some fashion to broaden the executive's power in the war-making area.

The available record of the remainder of the debate indicates, however, that the delegates may not have understood the change in such an unambiguous way, or indeed in any one way at all. For example, Roger Sherman of Connecticut protested that the clause as originally reported, that is, without the Madison-Gerry amendment, "stood very well. The Executive shd. be able to repel and not to commence war. 'Make' better than 'declare' the latter narrowing the power too much." It thus appears that Sherman believed that the original wording already left the executive free to repel sudden attacks and hence that in this respect the proposed change was nugatory. But he apparently thought that by narrowing the power of Congress, the alteration would unduly broaden the executive's power in some other ways. Oliver Ellsworth, also of Connecticut, at first concurred with Sherman in opposing the change. George Mason of Virginia, who "was against giving the power of war to the Executive, because [he was] not safely to be trusted with it; or to the Senate, because [it was] not so constructed to be entitled to it . . . and was for clogging rather than facilitating war," and who on that basis might well have agreed with Sherman and Ellsworth, instead supported the change. Madison's notes indicate that at this point a vote on the proposed change was taken, resulting in a tally of seven states to two in favor of the amendment. Then, in the face of an argument advanced by Rufus King of Massachusetts "that *'make'* war might be understood to 'conduct' it which was an Executive function," Ellsworth accepted the need for the alteration in wording and Connecticut switched its vote, making the poll eight to one in favor of the Madison-Gerry amendment.[16]

In editing the records of the Convention, Max Farrand concluded that the official journal of the Convention, as regards the

16. *Id.* at 318-19. *See also* Note, *Congress, the President, and the Power to Commit Forces to Combat, supra* note 4, at 1773 n.16. Only nine states cast votes because Massachusetts abstained on the issue, New Jersey and New York were not represented at this point, and Rhode Island never attended the Convention.

votes taken, was at times unreliable, inconclusive, or both,[17] and for this reason Madison's figures are used here. If accurate, these figures indicate that King's argument had but a marginal effect on the fate of the Madison-Gerry amendment, since it would have passed without Connecticut's vote. But this is a case where the journal figures do not accord with Madison's. The journal indicates that the Madison-Gerry amendment initially *lost* by a vote of four states to five and that it was only on the second vote that the amendment carried–by an eight to one margin.[18] *If* the journal is accurate in its voting figures and *if* King's argument was the key statement made between the two votes, then it had a far greater effect on the outcome than Madison's account would suggest.

The "ifs" must remain, for the record is unclear. The only certainty which emerges from the debate is that the wording was changed. What the various delegates thought the change accomplished can only be set forth in terms of possible interpretations: (1) the change broadened executive power by giving the executive the authority "to repel sudden attacks"; (2) the original wording already having given the executive that power, the alteration broadened his power in some more general fashion; (3) the modification lessened the chance for involvement in war; or (4) the new wording removed any suggestion that Congress would control the conduct of a war *after* it was begun. Of these possibilities, the first two suggest an intention to broaden the executive's power in *some* way. The third and, particularly the statement of Mason from which it is derived, does not bear clearly one way or the other on the question of executive-versus-legislative power. The fourth possibility suggests that the change referred to something other than the process of initially committing the nation to war. As if to reinforce the conclusion that the change meant different

17. 1 Farrand, Records, *supra* note 6, at xii-xiv. Madison himself later "corrected" some of his information to bring it into conformity with the journal, but such emendations were not made in his notes of the debate under discussion.

18. 2 *id.* at 313-14. Russell appears more convinced than I am that the journal's record of these votes is accurate, and he suggests that Madison may have mislabeled one of the vote counts. His position on the largely inconclusive meaning of the debate is, however, close to mine. *See* Russell, *supra* note 6, at 39-43.

things to different delegates, Butler of South Carolina, after the new wording had been approved, "moved to give the Legislature [the] power of peace as they were to have that of war."[19] In his mind, it seems, the Madison-Gerry amendment had done little. The legislature still had the power "of war."

II. Indications of an Original Understanding

If the discussion and votes on August 17 were the only source of evidence revealing the original understanding of war-making under the Constitution, those eager for a definite conclusion would have due cause for despair. But the private intentions of the Convention's delegates in changing "make" to "declare," whatever they were, did not control the way Americans of that day generally understood the Constitution's war-making clauses. Fortunately, we can gain additional and more conclusive evidence from (a) other deliberations of the Federal Convention and the Constitution itself; (b) the state debates over the ratification of the Constitution; (c) some seventeenth and eighteenth century trends in the theory and practice of war and reprisal; and (d) a consideration of English influences.

A. *The Convention and Its Product*

In 1787 and 1788, Americans ratified the Constitution, not a set of proceedings in the Federal Convention.[20] In fact, the Convention's records remained largely secret for thirty years.[21] Still, the deliberations in Philadelphia are worth examining for clues as to how Americans of that day generally understood war-making in its constitutional dimensions. The Convention's members were, after all, members of and leaders in a broader community, so it is likely that their fundamental assumptions were shared by other Americans. In addition, the Constitution itself provides clues to

19. 2 Farrand, Records, *supra* note 6, at 319.

20. *See, e.g.,* 1 J. Story, Commentaries on the Constitution of the United States 388-90 (1833).

21. 1 Farrand, Records, *supra* note 6, at xi-xv.

how the delegates' contemporaries understood war-making under it.

The plan of government submitted by Edmund Randolph on behalf of Virginia became the focus of the Convention's first deliberations. It contained no specific reference, however, to the power of either the executive or the legislative branch to commit the nation to war,[22] but instead neatly side-stepped the question. Randolph recommended that the "National Legislature . . . ought to be impowered to enjoy [among other things] the Legislative Rights vested in Congress by the Confederation" and that the "National Executive . . . ought to enjoy the Executive rights vested in Congress by the Confederation."[23] The only recorded remarks bearing on how such criteria for division might affect the allocation of war-making power came on June 1, 1787. Charles Pinckney cautioned that the powers of war and peace might properly be classed as executive powers. James Wilson of Pennsylvania, although admitting his preference for "a single magistrate, as giving most energy dispatch and responsibility to the office [of National Executive]," took exception: "[h]e did not consider the Prerogatives of the British Monarch [which included the power of making war] as a proper guide in defining the Executive Powers. Some of these prerogatives were of a Legislative nature. Among others that of war & peace &c." Madison agreed with Wilson,[24] but the resolutions which the Convention eventually passed and sent to the Committee on Detail on July 26 were no more explicit regarding the division of war-making power than Randolph's original plan had been. In a sense they were even less explicit because they did not contain the general proposition that the executive should enjoy the executive powers vested in the Confederation Congress.[25]

Despite the paucity of prior debate and the ambiguity of the resolutions sent to it, the Committee on Detail had little trouble in allocating the war-making power. Randolph and Wilson each pre-

22. *See id.* at 18-23.
23. *Id.* at 21.
24. *Id.* at 64-65, 70.
25. *See* 2 *id.* at 129-34.

pared draft constitutions which assigned the power "to make war" to the legislature. The draft reported by the committee to the Convention on August 6 followed the same scheme. Clearly, as the committee sensed the will of the Convention on these points—points which, it must be remembered, had scarcely been debated—war-making fell almost automatically to Congress. At the same time, the committee made the executive, now denominated the President, the Commander in Chief of the armed forces. In view of the concurrent grant to Congress of the broad power "to make war," the Presidency did not carry with it any authority to initiate war, except perhaps the restricted power of repelling sudden attacks which Sherman was soon to attribute to it. After the committee reported, the Convention spent a month debating and sometimes modifying its recommendations, changing, *inter alia,* Congress' power "to make war" to a power "to declare war." The Commander-in-Chief clause, however, was passed unchanged and without recorded debate on August 27, 1787. This expeditious, unremarked assent again suggests a narrow, non-controversial conception of the clause.[26]

Charles Pinckney's recommendation that Congress be given the power to grant letters of marque and reprisal was approved on September 5. Perhaps, as Joseph Story later contended, the Convention desired to remove any remaining doubt about the authority of Congress to authorize some form of undeclared hostilities.[27] At the least, it thus became possible for Americans in 1787-88 to draw such a conclusion, with their precise interpretation of the scope of the power depending on how they understood the purpose of letters of marque and reprisal.[28] The Convention's final

26. These developments can be traced in *id.* at 143, 145, 168, 172, 182, 185, 426-28; p. 8 *supra.* On the narrow view the Convention took of the President's role as Commander in Chief, see especially May, *"The President Shall Be Commander in Chief" (1787-1789)*, in THE ULTIMATE DECISION: THE PRESIDENT AS COMMANDER IN CHIEF 3-19 (E. May ed. 1960).

27. *See* 2 FARRAND, RECORDS, *supra* note 6, at 326, 508-09; 3 STORY, *supra* note 20, at 63-64.

28. Despite Justice Story's comment, recent students of the war-making issue have generally neglected the significance of the power to grant letters of marque and reprisal, which is discussed at pp. 27-32 *infra.* For an example of the complete omission of this power from a list of "specific powers relevant to [the] discussion" of the original understanding, see Rogers, *Congress, the President, and*

product contained all three provisions: Congress received the power to declare war and to grant letters of marque and reprisal, and the President became Commander in Chief.[29]

The Convention also left several less direct clues to how the delegates and their contemporaries may have understood the Constitution's war-making provisions. One is found in the plan of government that Alexander Hamilton presented to the Convention on June 18. This discloses that Hamilton, despite his preference for a greatly strengthened executive, was inclined to limit severely the executive's role in initiating war. He would have given the Senate, not the President, "the sole power of declaring war." The President was instead "to have the direction of war *when authorized or begun.*"[30] His scheme also supports the conclusion that the delegates and their contemporaries in America did not understand the term "declare" in a narrow, technical sense. On Hamilton's theory the President could direct war only after it had been commenced. His sole reference to the commencement of war, however, was the grant empowering the Senate to declare it–yet not even in the eighteenth century were all wars technically "declared."[31]

---

the War Powers, 59 CALIF. L. REV. 1194, 1195 (1971). Secretary of State Rogers' article appeared earlier as a prepared statement in *1971 Hearings on War Powers Legislation, supra* note 4, at 122.

29. U.S. CONST. art. I, § 8, art. II, § 2.

30. 1 FARRAND, RECORDS, *supra* note 6, at 292 (emphasis added). Notes on the plan Hamilton presented to the convention on June 18 are found in *id.* at 282-93. The version he presented in draft form to Madison late in the convention is in 3 *id.* at 617-30. On Hamilton generally during the Convention, see B. MITCHELL, ALEXANDER HAMILTON: YOUTH TO MATURITY 1755-1788, at 389-413 (1957); J. MILLER, ALEXANDER HAMILTON AND THE GROWTH OF THE NEW NATION 151-83 (1964).

31. During the debate of August 17, on the war-making clause, Elbridge Gerry (or perhaps Madison in recording Gerry's remarks) had used "declare" in a similarly loose sense. Pierce Butler proposed "vesting the power [*i.e.,* the power to *make* war, since the amendment to change the wording had not yet been proposed] in the President." Gerry, commenting after Madison's and his amendment to change the wording had been offered but before it had been voted on, and with evident reference to Butler's remark, soon objected that he "never expected to hear in a republic a motion to empower the Executive alone to *declare* war." 2 FARRAND, RECORDS, *supra* note 6, at 318 (emphasis added). Whether the precise wording in Madison's notes is an accurate rendition of Gerry's usage or whether it reflects Madison's imprint is immaterial. Either alternative supports the conclu-

Another clue is provided by the placement within the Constitution of the restrictions on the power of the states to wage war independently of the central government. In Philadelphia, limitations even stricter than those which had been included in the Articles[32] emerged during the deliberations of the Committee on Detail and were included as separate coordinate articles in the draft reported by that committee to the Convention. The Committee on Style, however, placed the prohibitions in the *legislative* article, where they appear in the final Constitution.[33] If the Committee on Style decided on this arrangement by reasoning that any authority possessed by the states to make war would derogate from the power of the new Congress, the committee assumed something that the Convention had never explicitly resolved—namely, that commencing war is properly a legislative function. Like the earlier action by the Committee on Detail in assigning the power "to make war" to Congress, the Committee on Style's action suggests that there was no need for clarification by the Convention, since the notion was generally accepted.

Two points may detract from the soundness of this conclusion. First, the limitations on states placed in Article I included restrictions on the foreign relations power of the states.[34] This would imply, using a similar argument based on placement, that the delegates also viewed foreign relations as properly within the legislative sphere, despite the apparent fact that the President was given substantial power in the foreign relations area. Offsetting the contention that on this account the argument based on placement is unsound is the fact that the Convention did provide the Senate with a check on the President's treaty-making powers, a check

---

sion that "declare" did not have a very strict meaning in current American usage. This conclusion is elaborated at p. 18, pp. 28-32 *infra*.

32. *See* note 10 *supra*.

33. *See* 2 FARRAND, RECORDS, *supra* note 6, at 169, 187, 577, 597; U.S. CONST. art. I, § 10: "No State shall . . . grant letters of Marque and Reprisal . . . [nor,] without the Consent of Congress, . . . engage in War, unless actually invaded, or in such imminent Danger as will not admit of delay."

34. "No State shall enter into any Treaty, Alliance, or Confederation . . . [nor,] without the Consent of Congress, . . . enter into any Agreement or Compact with another State, or with a foreign Power . . . ." U.S. CONST. art. I, § 10. Neither here nor in the restrictions on state war-making did the final document contain any substantive changes from the report of the Committee on Style.

that George Washington at first took especially seriously.[35] In addition, it gave Congress far more power than the President in what contemporaries hoped and thought would be the dominant and proper form of America's relations with the world–commercial relations.[36] Similarly detracting from the force of the argument based on placement is the fact that we simply do not know much about the considerations which guided the Committee on Style.[37] It is quite possible that in placing the restrictions on state war-making, the committee and its chief draftsman, Gouveneur Morris of Pennsylvania, were not concerned with the implications just discussed. Yet the suspicion remains that something more than style prompted the committee's change, since the limitations could easily have been continued as one or more separate articles or else placed in Article IV, which contains other provisions relating to the states. In any event, placement of the state war-making restrictions may have contributed to a general impression that Congress was intended to be dominant in the field.

A final indirect clue to how Americans in the late 1780's understood war-making under the Constitution comes from the provision that states may not engage in war "unless actually invaded, or in such imminent Danger as will not admit of delay."[38] Although the Convention had given some attention to the issue of surprise attack during its one debate over the main war-making grant to Congress,[39] this is the only explicit reference to the matter in either the completed Constitution or the earlier drafts. While the provision may appear inconsequential to a late twentieth century observer, it undoubtedly had far more meaning in the 1780's

35. *See* R. Hayden, The Senate and Treaties 1789-1817, at 1-106 (1920). Hayden concludes that "Washington made treaties 'by and with the advice and consent' of the Senate in a sense and to an extent that no later President ever has." *Id.* at 103.

36. *See, e.g.*, P. Varg, Foreign Policies of the Founding Fathers 1-69 (Penguin ed. 1970); G. Stourzh, Benjamin Franklin and American Foreign Policy 238-46 (2d ed. 1969); F. Gilbert, The Beginnings of American Foreign Policy: To The Farewell Address *passim* (1965).

37. *See* Letter from James Madison to Jared Sparks, April 8, 1831, in 3 Farrand, Records, *supra* note 6, at 498, 499; C. Rossiter, 1787: The Grand Convention 224-30 (1966); M. Farrand, The Framing of the Constitution of the United States 181-82 (1913).

38. U.S. Const. art. I, § 10.

39. *See* pp. 7-10 *supra*.

when communications and transportation would not have allowed an immediate federal response to a truly surprise attack. In such a situation, the real problem would have been whether states might act prior to a national decision. The provision allowing state action may have seemed more consequential, too, both because state-versus-federal authority was a major issue in the general constitutional controversy of the day, and because the Constitution otherwise narrowed the sphere of permissible state activity in the war-making area. Yet the completed Constitution contained no indication that the states were to have exclusive responsibility to meet surprise attack.[40] In sum, the question of how an observer in the late 1780's would have interpreted the provision cannot be definitely answered, but its presence in the Constitution at least further suggests that Americans of that day need not have envisaged that the President as Commander in Chief would have an especially broad role in repelling sudden attack.

B. *The State Ratification Debates*

The state ratification debates, like the Philadelphia Convention, were little concerned with how the new government would initiate war. As an example, five of the eleven states which ratified the Constitution before the new government commenced operation offered amendments to it, but of the seventy-seven amendments thus proposed, only one–from New York–dealt with the power of Congress to declare war. Moreover, that proposal, to require a two-thirds vote in each house of Congress for a declaration of war, was designed to protect state or regional interests rather than to alter the balance of war-making power between Congress and the President. By contrast, twenty-eight of the proposed amendments concerned elections and procedures under the new government, fourteen dealt with individual rights and privileges, nine with taxation and finance, six with the raising and maintenance of armies and control of the militia (*i.e.,* the war-*supporting* function) and five each with commerce and the jurisdiction of courts.[41]

40. *See also* p. 21 *infra.*
41. *See* Ames, *The Proposed Amendments to the Constitution of the United*

Contemporary newspaper, pamphlet, and state convention debates display a similar lack of attention to the allocation of the war-making power. Although, for example, the first North Carolina convention eventually found the Constitution objectionable enough not to ratify it, the clause giving Congress the power "to declare War [and] grant Letters of Marque and Reprisal" was "read without any observation."[42] James Winthrop in Massachusetts and Richard Harry Lee in Virginia both objected to the increased centralization which characterized the proposed government, but both agreed that the power of war could, in Lee's words, "be lodged no where else, with any propriety, but in this [the central] government."[43] The explicit restrictions on state war-making in the Constitution received almost no attention, adverse or otherwise, in the state debates.[44] Though the authors of *The Federalist Papers* were not above beating a straw man in arguing their case,[45] in Number 41 Madison commented: "Is the power of

---

States During the First Century of Its History, in 2 ANNUAL REPORT OF THE AMERICAN HISTORICAL ASSOCIATION FOR THE YEAR 1896, at 307-09 (1897). For the texts of the amendments, see 1 DEBATES IN THE SEVERAL STATE CONVENTIONS ON THE ADOPTION OF THE FEDERAL CONSTITUTION 322-31 (J. Elliot ed. 1888); 3 *id.* at 659-61 [hereinafter cited as ELLIOT, DEBATES]. The five states offering amendments, in order of ratification, were Massachusetts, South Carolina, New Hampshire, Virginia, and New York. The amendments nearly proposed by Maryland, those declared by the first North Carolina convention to be necessary before that state ratified, and those proposed by Rhode Island in its tardy ratification also support the conclusions in the text. *See* Ames, *supra,* at 309-10; 1 ELLIOT, DEBATES 336-37; 2 *id.* at 550-53; 4 *id.* at 244-47. In several instances my classification of an amendment differs from that of Ames. In any event, only the New York amendment noted in the text and another New York amendment to prohibit the President as Commander in Chief from commanding the Armed Forces in person (both of which are in 1 *id.* at 330) bore on the external war-making powers of the proposed government.

42. 4 *id.* at 94.

43. ESSAYS ON THE CONSTITUTION OF THE UNITED STATES 1787-1788, at 98 (P. Ford ed. 1892) [hereinafter cited as FORD, ESSAYS]; PAMPHLETS ON THE CONSTITUTION OF THE UNITED STATES 1787-1788, at 300-01 (P. Ford ed. 1888) [hereinafter cited as FORD, PAMPHLETS].

44. "Almost," because Madison made cursory mention of them. *See* THE FEDERALIST No. 44 (J. Cooke ed. 1961) (except as noted, all citations herein to *The Federalist* are to the definitive Cooke edition).

45. Publius argued that opponents of the Constitution were proposing a series of regional confederations in place of a federal union, but the latest study of the ratification controversy in New York finds practically no talk of this sort among

declaring war necessary? No man will answer this question in the negative. It would be superfluous, therefore, to enter into a proof of the affirmative."[46] The existing Confederation government already held the power to make war. Its presence in the Constitution could hardly be a matter of controversy. In addition, the war-making power did not have the direct connection with such broader issues as state-versus-federal taxation and civilian-versus-military rule that the war-*supporting* powers had—the latter connection producing considerable debate.[47]

Yet, again like the Philadelphia Convention, the state debates offer revealing, albeit indirect and sometimes inferential, evidence about how contemporaries understood the Constitution's war-making clauses. Significantly, several comments strongly hint that Americans in 1787-88 thought the power to *declare* war assigned to the new Congress was practically identical with the old Congress' power of *determining on war.* After stating in *The Federalist Number 41* that it would be superfluous to examine whether the power of declaring war was necessary, Madison remarked: "The existing confederation establishes this power in the most ample form."[48] In the New York convention John Jay implicitly equated the power of the old Congress to determine on war with the power to declare war. Robert R. Livingston was more direct: "But, say the gentlemen, our present [Confederation] Congress have not the same powers [as the new Congress]. I answer, They have the very

---

the state's antifederalists. *See* THE FEDERALIST Nos. 3, 5 (J. Jay), No. 13 (A. Hamilton); L. DEPAUW, THE ELEVENTH PILLAR: NEW YORK STATE AND THE FEDERAL CONSTITUTION 173 (1966).

46. THE FEDERALIST No. 41, at 269-70. I have sought to avoid over-reliance on *The Federalist* as a guide to the original understanding. As Professor McLaughlin remarked years ago, "[T]hese essays were probably of service in winning support of the Constitution; but the extent of that service we naturally cannot measure. For much immediate practical effect they were perhaps too learned, too free from passion. . . . *The Federalist* probably had more effect after the new government went into operation than in the days of uncertainty when the fate of the union seemed to hang in the balance. . . ." A. MCLAUGHLIN, A CONSTITUTIONAL HISTORY OF THE UNITED STATES 208-09 (1935).

47. *See* Donahoe & Smelser, *supra* note 5, and the citations contained therein; A. EKIRCH, THE CIVILIAN AND THE MILITARY 27-31 (1956).

48. THE FEDERALIST No. 41, at 270. Madison reiterated this view in the Virginia convention. *See* 3 ELLIOT, DEBATES, *supra* note 41, at 259.

same . . . [including] the power of making war . . . ."[49] During the Pennsylvania convention James Wilson not only implictly equated declaring war and entering war, but also explicitly foreclosed exercise of the power by the President acting alone:

> This [new] system will not hurry us into war; it is calculated to guard against it. It will not be in the power of a single man, or a single body of men, to involve us in such distress; for the important power of declaring war is vested in the legislature at large; this declaration must be made with the concurrence of the House of Representatives: from this circumstance we may draw a certain conclusion that nothing but our national interest can draw us into a war.[50]

Consistent with these comments indicating a broad view of the power of the new Congress, supporters of the Constitution described the position of the President as Commander in Chief in narrow terms. Most notably, Hamilton contended that the office involved only command in a military sense, with no policy role. The President's authority, he wrote,

> would be nominally the same with that of the King of Great Britain, but in substance much inferior to it. It would amount to nothing more than the supreme command and direction of the military and naval forces, as first general and admiral of the confederacy; while that of the British king extends to the *declaring* of war, and to the *raising* and *regulating* of fleets and armies; all which by the constitution under consideration would appertain to the Legislature.[51]

James Iredell, who contended that "[o]ne of the great advantages attending a single Executive power is the degree of secrecy and dispatch with which on critical occasions such a power can act,"[52] nevertheless described the office of Commander in Chief in a constrained fashion strikingly similar to Hamilton's portrayal.[53]

49. 2 *id.* at 278, 284.
50. *Id.* at 528.
51. The Federalist No. 69, at 465.
52. Ford, Pamphlets, *supra* note 43, at 352.
53. *See* 4 Elliot, Debates, *supra* note 41, at 107-08. Iredell was not inconsistent in these two positions, for his quotation in this paragraph comes from a discussion of presidential conduct *during* war.

In accordance with this view, the federalists ignored a clear opportunity to describe the office of Commander in Chief in broad terms. At issue was the fear harbored by contemporaries that joining the purse with the sword would promote tyranny. There was a check in this regard even in England, where, as Patrick Henry observed, "The King declares war; the House of Commons gives the means of carrying it on."[54] The Constitution, though, would join the two powers in the new central government or, according to a noteworthy variation of the argument, in Congress.[55] The antifederalists, of course, preferred to remedy the situation by giving the states greater control over taxation and the raising of armies—a solution obviously unacceptable to the Constitution's supporters, who, however, were not themselves united on the issue. Hamilton and Madison defended the proposed arrangements in part by claiming that the Constitution did separate the purse and the sword, with Congress holding the one and the President the other; but on balance they tended not to emphasize this division. Instead, to meet the antifederalist argument, Hamilton and Madison drew attention to the need for federal control of the purse and thereby avoided stressing a broad role for the President as Commander in Chief.[56] Other federalists, showing even more reluctance to take an expansive view of the executive's power in this area, produced a different and revealing defense. "Are the people of England more secure," asked John Marshall, "if the Commons have no voice in declaring war? or are we less secure by having the Senate [*sic*] joined with the President?"[57] Oliver Ellsworth posed a similar question:

> [D]oes it follow, because it is dangerous to give the power of the sword and purse to an hereditary prince, who is independent of the people, that therefore it is dangerous to give it to the Parliament—to Congress, which is your Parliament—to men appointed by yourselves, and dependent upon yourselves? This argument

54. 3 *id.* at 172.

55. *See* 2 *id.* at 376-77; 3 *id.* at 172, 378-79. For the sources of the concern over the joining of purse and sword, see May, *supra* note 27; Ekirch, *supra* note 47, at 3-24; L. Smith, American Democracy and Military Power 20-24 (1951).

56. *See* 2 Elliot, Debates, *supra* note 41, at 348-51; 3 *id.* at 393-94.

57. *Id.* at 233.

amounts to this: you must cut a man in two in the middle, to prevent his hurting himself.[58]

When judged against future developments, "Publius" may have been "understandably wrong . . . in giving a purely military cast to the President's authority as commander-in-chief,"[59] but the evidence indicates that his view in this respect accorded well with that of his contemporaries in the state debates.

Even so, federalist comments about the desirability of governmental energy, efficiency, and dispatch are sometimes taken to indicate a latitudinarian view of the executive war-making power.[60] Indeed, during the ratification controversy, the federalists contended that the proposed national government's enhanced ability to raise armies and build fleets would promote national security, in part by deterring surprise attack on the United States.[61] That this was a "winning issue" suggests contemporaries were probably not convinced that the constitutional provision allowing state response to surprise attack[62] was an adequate safeguard by itself, which in turn gives color to the conclusion that they looked to the President to act in cases of sudden attack. But during the ratification debates, the federalists never in fact defended the presidential office on grounds that its energy and dispatch were required in the

58. 2 *id.* at 195. *Cf.* comments of Edmund Randolph, who had refused to sign the Constitution in Philadelphia and eventually voted for ratification in Virginia only after considerable wavering. 3 *id.* at 201.

59. Rossiter, *Introduction* to THE FEDERALIST PAPERS xiii (C. Rossiter ed. 1961). *But cf.* C. ROSSITER, THE SUPREME COURT AND THE COMMANDER IN CHIEF 67 (1951).

60. *See, e.g.,* Rogers, *supra* note 28, at 1196 & n.10, in which, however, the key statement of the argument (the text associated with footnote 10) is simply not supported by a reading of the authorities cited in that footnote. On the discrepancy, *compare id., with* THE FEDERALIST Nos. 49 & 63 (both Madison, but misassigned by Rogers to Hamilton through use of a nineteenth century edition of *The Federalist*), Nos. 70-75 (Hamilton). Although certain of these numbers of *The Federalist* discuss the conduct and direction of war (Nos. 70, 72, 74) and the presence of secrecy and dispatch in the Constitution's treaty-making process (*i.e.*, including the Senate) (No. 75), none discusses the topics of the *commencement* of war or other hostilities against another nation, or the locus-of-power problem with respect to such commencement.

61. *See* Marks, *Foreign Affairs: A Winning Issue in the Campaign for Ratification of the United States Constitution,* 86 POL. SCI. Q. 444 (1971).

62. *See* pp. 15-16 *supra.*

commencement of war against a foreign enemy in emergency situations. What emerges instead is the conclusion that they were intent on defending the national government without regard to a particular branch.[63]

In a similar vein, although the federalists undoubtedly envisioned the Constitution as designed both to meet immediate problems and to comprehend future demands, they seem not to have discussed the allocation of power to commence war in the context of such a theory. Writing in *The Federalist Papers,* for example, Hamilton boldly noted, "The authorities essential to the care of the common defence . . . ought to exist without limitation: *Because it is impossible to foresee or define the extent and variety of national exigencies, or the correspondent extent & variety of the means which may be necessary to satisfy them.*"[64] He subsequently reiterated the point: "There ought to be a capacity to provide for future contingencies, as they may happen; and as these are illimitable in their nature, so it is impossible safely to limit that capacity."[65] In the first instance, however, he was specifically concerned with the powers "to raise armies–to build and equip fleets–to prescribe rules for the government of both–to direct their operations–to provide for their support." In the second, he was considering the problem of providing an adequate revenue for the proposed government. In sum, although defending the notion of a flexible, expandable Constitution, these statements plainly refer to immediate problems other than that of commencing war. Moreover, Hamilton's broad purpose in both instances was to defend federal power, *once more without specification of branch,* against exceptions and reservations in favor of the states.[66]

Of course, their silence does not necessarily mean Americans in the late 1780's rejected the idea that the President had responsibility to respond to sudden attack. Particularly in view of the common expectation that George Washington would be the first President,[67] it is conceivable that they simply and tacitly assumed that

63. *See* Marks, *supra* note 61, at 456-57, and the citations therein.

64. The Federalist No. 23, at 147.

65. *Id.* No. 34, at 211.

66. *See id.* Nos. 23, 34. *See generally id.*, Nos. 23-36 (A. Hamilton).

67. *See, e.g.*, 6 D. Freeman, George Washington: A Biography 117 & n.1 (1954).

there would be a presidential role in this regard. There was certainly sentiment present in the Federal Convention that "[t]he Executive shd. be able to repel . . . war" as Roger Sherman had explained.[68] What the preponderance of evidence suggests, however, is that *if* men of the day generally shared such an assumption, they still conceived of the President's war-making role in exceptionally narrow terms.

C. *The Theory and Practice of War and Reprisal*

Americans of the revolutionary generation paid considerable attention to a broad range of European and especially English ideas and controversies involving law, government, and international affairs.[69] The works of Hugo Grotius,[70] Samuel Pufendorf,[71] Emmerich de Vattel,[72] and particularly Jean Jacques Burlamaqui[73] were widely read and quoted.[74] Books by Thomas Rutherforth[75]

68. *See* pp. 7-10 *supra.*

69. *See, e.g.,* B. BAILYN, THE IDEOLOGICAL ORIGINS OF THE AMERICAN REVOLUTION 22-54 (1967); 1 A. CHROUST, THE RISE OF THE LEGAL PROFESSION IN AMERICA: THE COLONIAL EXPERIENCE 33-37 (1965); T. COLBOURN, THE LAMP OF EXPERIENCE: WHIG HISTORY AND THE INTELLECTUAL ORIGINS OF THE AMERICAN REVOLUTION (1965); GILBERT, *supra* note 35; M. KRAUS, THE ATLANTIC CIVILIZATION: EIGHTEENTH CENTURY ORIGINS (1949); Bailyn, *Political Experience and Enlightenment Ideas in Eighteenth-Century America,* 67 AM. HIST. REV. 339 (1962).

70. H. GROTIUS, THE RIGHTS OF WAR AND PEACE (1625; J. Barbeyrac trans. 1738) (to facilitate reference, here and in the following notes I have indicated both an early edition and the edition I used) [hereinafter cited as GROTIUS].

71. S. PUFENDORF, ON THE LAW OF NATURE AND NATIONS (1688; C. & W. Oldfather transl. 1934) [hereinafter cited as PUFENDORF].

72. E. DE VATTEL, THE LAW OF NATIONS (1758; 2 vols. in 1, trans. from French 1759-60) [hereinafter cited as VATTEL].

73. J. BURLAMAQUI, THE PRINCIPLES OF NATURAL AND POLITIC LAW (2 vols., T. Nugent trans. 1752; 3rd ed. 1784) [hereinafter cited as BURLAMAQUI].

74. *See* BAILYN, *supra* note 69, at 27-29; P. HAMLIN, LEGAL EDUCATION IN COLONIAL NEW YORK 197-99 (1939); R. HARVEY, JEAN JACQUES BURLAMAQUI: A LIBERAL TRADITION IN AMERICAN CONSTITUTIONALISM 79-105 (1937); C. WARREN, A HISTORY OF THE AMERICAN BAR 163, 181-82 (1911); B. WRIGHT, AMERICAN INTERPRETATIONS OF NATURAL LAW 7-8, 44, 50-51, 58, 60, 67, 79, 89-90 (1931); Weinfeld, *What Did the Framers of the Federal Constitution Mean by "Agreements and Compacts"?,* 3 U. CHI. L. REV. 453, 458-59 (1936); cases cited in note 76 *infra.* BAILYN, *supra,* at 28, cautions, however, that the Americans' knowledge of these works was sometimes superficial; but this was probably less true of the lawyers who used them.

75. T. RUTHERFORTH, INSTITUTES OF NATURAL LAW (2 vols. 1754-56; Am. Ed. 1799) (also spelled Rutherford) [hereinafter cited as RUTHERFORTH].

and Richard Lee[76] (whose work was largely a popularization of the views of Cornelius van Bynkershoek)[77] appear to have been less widely read but still received attention.[78] Many of the fairly well educated and cosmopolitan men to whom the deferential American society of that day looked for leadership[79] thus must have been familiar with these treatises. Among the topics the treatises considered were several which helped illuminate the war-making power of the new government being considered in 1787-88. Consequently, these works, in conjunction with historical trends

76. R. LEE, A TREATISE OF CAPTURES IN WAR (1759) [hereinafter cited as LEE].

77. C. VAN BYNKERSHOEK, A TREATISE ON THE LAW OF WAR: BEING THE FIRST BOOK OF HIS QUAESTIONUM JURIS PUBLICI (1737; P. DuPonceau trans. 1810) [hereinafter cited as BYNKERSHOEK].

78. In November 1782 the Confederation Congress appointed a committee to compile a list of books to be purchased for its use. In January 1783 the committee reported a list of 309 books, only to have the budget-minded Congress decline to buy them. The list, which is indicative of recommended reading for the statesman of the 1780's and of which James Madison was probably the principal author, included not only Lee and Rutherforth but also Grotius, Pufendorf, Vattel, and Burlamaqui. Lee was in Thomas Jefferson's library for the period. Luther Martin read portions of Rutherforth to the Constitutional Convention (Farrand in the index to his RECORDS lists this as Samuel Rutherford, but Madison's and Yates' notes contain no indication of a first name, Thomas Rutherforth's name was spelled both Rutherford and Rutherforth, and the context would suggest it was Thomas Rutherforth's work which was being read); Hamilton quoted Rutherforth in *Federalist* 84; and James Wilson cited Rutherforth in his 1790-91 law lectures. Bynkershoek's *Quaestionum* appears to have been much less well-known, with the 1783 "statesman's" list merely giving the title of another book of his, followed by "with all his other works." Within several years of the Constitution's framing and ratification, Lee, Rutherforth, and Bynkershoek were all extensively cited by counsel in a number of prize cases heard by the U.S. Supreme Court. *See* L. Smith, The Library List of 1783, 1969, at 2-3, 123-25, 127, 129-30 (unpublished Ph.D. dissertation in the Honnold Library of the Claremont Colleges); 2 CATALOGUE OF THE LIBRARY OF THOMAS JEFFERSON 74 (E. Sowerby comp. 1953); 1 FARRAND, RECORDS, *supra* note 6, at 438, 440; 4 *id.* at 209; 1 THE WORKS OF JAMES WILSON 163, 192 (R. McCloskey ed. 1967); THE FEDERALIST No. 84; Glass v. The Sloop Betsey, 3 U.S. (3 Dall.) 6 (1794); Talbot v. Jansen, 3 U.S. (3 Dall.) 133 (1795); M'Donough v. Dannery and the Ship Mary Ford, 3 U.S. (3 Dall.) 188 (1796); Bas v. Tingy, 4 U.S. (4 Dall.) 37 (1800).

79. *See* Pole, *Historians and the Problem of Early American Democracy,* 67 AM. HIST. REV. 626 (1962). In saying that American society of the 1780's was deferential, I do not intend to become involved in the much-debated question of whether it was "democratic," politically or otherwise. Deference and democracy are not necessarily mutually exclusive.

visible in the late eighteenth century, offer further insight into the original understanding of war-making under the Constitution, and particularly the problem of undeclared war.

The writers in question discussed the necessity of declarations of war, but reached no consensus. Although none of them held a declaration necessary in a defensive war, Grotius, Pufendorf, and Vattel thought one required in a non-defensive war, if that war was to be legal with respect to the consequences attaching to it (for example, immunity from criminal prosecution for those waging it). Burlamaqui, however, was not entirely clear on the point. He stated that in a non-defensive war a declaration "ought" to be issued as a token of the respect which sovereigns should show each other, and so left unclear whether there was here a legal obligation or only some principle of comity. The remaining authors contended that a declaration was never legally necessary, but they agreed with Burlamaqui that it was desirable in that it put neutrals on notice to observe their obligations toward the belligerents and allowed the enemy one last chance to give satisfaction. Several of the writers stated that a declaration might be either absolute or conditional. The absolute declaration was a simple and unconditional announcement of war. The conditional declaration was essentially an ultimatum, that is, a demand that the enemy perform some act, coupled with a warning that otherwise war would follow as a matter of course with no further notice.[80]

It must initially be recognized that when these authors spoke of declarations of war, they generally meant formal announcements to the enemy with proper ceremony, usually in his capital.[81] In point of fact, however, European states had not made such declarations for well over a century.[82] The common practice was to publish an announcement domestically and to inform the enemy, if at all, by means of a written communication such as a diplo-

80. *See* Grotius 57-58, 552-55; Pufendorf 1294-95, 1307 (Pufendorf mainly refers his readers to Grotius on the subjects of declarations of war and reprisals); 2 Vattel 3, 21-26; 2 Burlamaqui 269-72; Bynkershoek 6-27; Lee 13-39; 1 Rutherforth 451; 2 *id.* at 522-23, 549-53.

81. *See* citations in note 80 *supra.*

82. This may have been what Hamilton had in mind when he wrote that "the ceremony of a formal denunciation of war has of late fallen into disuse . . . ." The Federalist No. 25, at 161.

matic note.[83] To cope with the disparity between theory and practice, Burlamaqui had sought to distinguish between a "declaration" and a "publication" of war.[84] But, with the possible exception of two American treaties concluded during the revolutionary and Confederation periods,[85] there appears to have been no acceptance of this distinction in the United States.[86] In view of these facts, it seems probable that a contemporary would have taken "declaration of war," as used in the treatises, to mean what nations had *customarily* done in "declaring" war during the preceding century or so.[87]

Once that equation was made, the treatises may well have affected the way those who debated the Constitution in 1787-88 understood the document. One intriguing possibility is that a person familiar with the literature on conditional declarations of war could easily have concluded that Congress might conditionally authorize the President to conduct a war if another nation refused

83. *See* G. MARTENS, SUMMARY OF THE LAW OF NATIONS FOUNDED IN THE TREATIES AND CUSTOMS OF THE MODERN NATIONS OF EUROPE 274-75 (1788; W. Cobbett trans. 1795); 2 J. WESTLAKE, INTERNATIONAL LAW 20-21 (2d ed. 1913).

84. 2 BURLAMAQUI 272-73. Vattel hinted at the same distinction (*see* 2 VATTEL 25), but generally used "declaration" to include both formal denunciation to the enemy and publication.

85. In their English (*i.e.,* American) texts, the Treaties of Amity and Commerce with France, Feb. 6, 1778, 8 Stat. 12 (1778), T.S. No. 83, and the Netherlands, Oct. 8, 1782, 8 Stat. 32 (1782), T.S. No. 249, used the phrase "proclamation of war" in articles (22 and 18, respectively) relating to the effects of an announcement of war *within* the announcing nation and used "declaration of war" elsewhere. The French and Dutch texts did not observe the distinction. *See* 8 Stat. 12, 32 (1778, 1782), 2 T.I.A.S., 2 TREATIES AND OTHER INTERNATIONAL ACTS OF THE UNITED STATES OF AMERICA 3, 59 (H. Miller ed. 1931). The original instructions of the Continental Congress regarding the French treaty (the so-called Treaty Plan of 1776) also included the distinction. *See* 5 JOURNALS OF THE CONTINENTAL CONGRESS 1774-1789, at 774 (1906). Other contemporary American treaties ignored it.

86. *See, e.g.,* Treaty of Amity and Commerce with Sweden, Apirl 3, 1783, 8 Stat. 60 (1783), T.S. No. 346, 2 TREATIES AND OTHER INTERNATIONAL ACTS, *supra* note 85, at 123; Convention of Peace, Commerce, and Navigation with France, Sept. 30, 1800, 8 Stat. 178 T.S. No. 85, 2 TREATIES AND OTHER INTERNATIONAL ACTS, *supra,* at 457. These used "declaration of war" in contexts similar to those in which the treaties cited in note 85 *supra* used "proclamation of war."

87. That at least some writers in the general period did so is seen in J. KENT, DISSERTATIONS, BEING THE PRELIMINARY PART OF A COURSE OF LAW LECTURES 66 (1795); MARTENS, *supra* note 83, at 274-75.

to meet American demands.[88] More importantly, though, the treatise writers' contention that declarations were not needed in defensive wars raised the question of undeclared war, and this was a matter that the treatises considered at some length.

Grotius had argued that declared war was "perfect" war in the sense of being complete—that is, involving whole nations on each side. Accordingly, undeclared war was "imperfect." The latter occurred "where no perfect War is absolutely denounced; yet where a certain violent protection of our rights is necessary," with the violence consisting of state-authorized private reprisals directed at property held by the subjects of another nation.[89] The later commentators agreed that imperfect war and reprisals were closely related, if not identical, but they took a broader view of what constituted reprisals, holding that states using public forces might themselves make reprisals. Burlamaqui's view is representative:

> A perfect war is that which intirely interrupts the tranquillity of the state, and lays a foundation for all possible acts of hostility. An imperfect war, on the contrary, is that which does not intirely interrupt the peace, but only in certain particulars, the public tranquillity being in other respects undisturbed.
>
> . . . This last species of war is generally called reprisals, of the nature which we shall here give some account. By reprisals then we mean *that imperfect kind of war, or those acts of hostility which sovereigns exercise against each other, or, with their consent, their subjects, by seizing the persons or effects of the subjects of a foreign commonwealth, that refuseth to do us justice . . .* .[90]

88. It would probably be anachronistic to equate such a view with permissiveness toward *delegation of power* in the modern sense.

89. *See* GROTIUS 538-49 (the quotation is at 540). *See also id.* at 57-58.

90. 2 BURLAMAQUI 258. *See* LEE 20, 40-51; 2 RUTHERFORTH 485, 516, 522-23, 537-53; 1 VATTEL 249-51; 2 *id.* at 25-26 (but Vattel's equation of imperfect war with reprisals is less completely developed than that of the other writers); 1 M. HALE, HISTORIA PLACITORUM CORONAE: THE HISTORY OF THE PLEAS OF THE CROWN 162 (S. Emlyn ed. 1736). Hale's comments in this regard are particularly significant, because, compared with other English jurists prior to Blackstone, "Hale was a particularly well-known and attractive figure" to Americans. BAILYN, *supra* note 69, at 30 n.11. It is also noteworthy that language practically identical to that in the first paragraph of the quotation from Burlamaqui was included, but without citation in Miller v. The Ship Resolution, 2 U.S. (2 Dall.)

There was consensus that imperfect war and state reprisals, commonly called general reprisals, could easily lead to perfect war.[91]

These opinions of the eighteenth century authors comported with two features of recent history, as seen in the late 1780's. First, hostilities without declarations of war were common during the period. Indeed, the undeclared hostilities in 1754-56 which marked the beginning of the Seven Years War between Britain and France, occurred mainly in America. During the War of the American Revolution, moreover, Britain and France never expressly declared war on each other, unless France's treaty with the United States may be so construed.[92] The second feature involved the practice of European states with respect to reprisals. From the twelfth through the seventeenth centuries, these states had regularized and legitimated private reprisals in time of peace by the sovereign's granting of letters of marque and reprisal to individuals who had specific claims against subjects of other states. Such letters authorized seizure of the property and sometimes the persons of the other state's subjects. When issued during war, the letters empowered individuals who were not members of public

---

1, 21 (Ct. App. in Cases of Capture 1781). Suggestive of the broadened concept of reprisals is this comment from an American about twenty years after the framing and ratification of the Constitution:

> Reprisals are either *general* or *special.*—They are *general* when a sovereign, who has or thinks that he has received an injury from another prince, issues orders to his military officers, and delivers commissions to his subjects to take the persons and property of the other nation, wherever the same may be found. It is, at present, the first step which is generally taken at the commencement of public war, and is considered as equivalent to a declaration of it.
>
> *Special* reprisals are granted, in time of peace, to individuals who have suffered an injury from the subjects of another nation . . . .

BYNKERSHOEK 182 n. * (trans. note). *See generally* 7 J. MOORE, A DIGEST OF INTERNATIONAL LAW 119-30 (1906).

91. *See* 2 BURLAMAQUI 261; LEE 20, 45; 1 VATTEL 250. *See also* GROTIUS 57-60. While Rutherforth does not explicitly deal with the point, neither does he contradict the conclusions of the other writers on it.

92. *See* J. MAURICE, HOSTILITIES WITHOUT DECLARATION OF WAR 1700-1870, at 12-26 (1883); 3 R. PHILLIMORE, COMMENTARIES ON INTERNATIONAL LAW 110-15 (1857); 2 WESTLAKE, *supra* note 83, at 22-23. Phillimore contends (at 115) that France's announcement of her American treaties in 1778 had the effect of a declaration.

armed forces to take from the enemy and his subjects. Then, during the first half of the eighteenth century, the practice of granting letters of marque and reprisal for satisfaction of private claims during peace virtually disappeared. States, however, continued to make reprisals to press their own claims, using both public naval forces and private ships sailing under privateer commissions or letters of marque and reprisal. These state or general reprisals not uncommonly resulted in outright war. English history furnished several such examples which were undoubtedly familiar to those who debated the Constitution: the wars with the Netherlands in 1652 and 1664, with Spain in 1739, and with France in 1756 were all preceded by public naval reprisals.[93]

In sum, familiarity with Grotius and his successors and with then-recent history would have suggested to one in the late 1780's that undeclared war was no oddity and that the issuance of letters of marque and reprisal for satisfaction of private claims was outmoded. An American who was knowledgeable about these topics therefore faced the problem of explaining the Constitution's use of the wording "declare" and "letters of marque and reprisal." He might have done this in several ways, but certain interpretations would have made more sense to him than others.

In regard to the word "declare," he might have settled on any of a variety of possibilities:

(1) *America would restrict herself to fully declared wars.* However, this interpretation would have run counter to the prevalence of undeclared war in the eighteenth century, which would have been a matter of some importance to a generation that made a practice of deriving lessons from history.[94] In addition, it would have contradicted a goal of the new diplomacy with which Amer-

93. *See* A. Hindmarsh, Force in Peace: Force Short of War in International Relations 43-56 (1933); 3 Phillimore, *supra* note 92, at 108-09, 118; Maccoby, *Reprisals as a Measure of Redress Short of War,* 2 Camb. L.J. 60, 60-67 (1924); 1 Letters and Papers Relating to the First Dutch War 1651-54, at 301-02 (S. Gardiner ed. 1899); 2 The Royal Navy: A History 422-25 (W. Clowes ed. 1898); 3 *id.* at 51-52, 266-67; 1 J. Corbett, England in the Seven Years War 83 (2d ed. 1918).

94. *See* Adair, *"Experience Must Be Our Only Guide": History, Democratic Theory, and the United States Constitution* in The Reinterpretation of Early American History 129 (R. Billington ed. 1966); Colbourn, *supra* note 69 *passim* and especially 4-6, 185.

icans were so enamored, namely, that force should be restricted and the effects of war controlled.[95] Indeed, in 1785 provisions reflecting this goal had been included in the Treaty of Amity and Commerce between the United States and Prussia.[96] But fully declared wars were less likely to be limited in their material dimensions.

(2) *The provision giving Congress the power to declare war was merely a formality. In practice, America would avoid war by limiting her relations with the world to trade and commerce.*[97] Many Americans undoubtedly hoped the United States could remain aloof from war, but a generation which avidly followed European affairs, which had confronted European activities in America and Barbary depredations in the Mediterranean during the 1780's,[98] and which contained some eminent realists[99] would hardly have counted on the complete disappearance of war.

(3) *Congress' power to declare war was to be interpreted strictly; as a result the Constitution left the waging of undeclared war unaccounted for. Authority over it must have been either unvested or perhaps somewhat deviously vested in the executive branch.* It seems unlikely, however, that an observer in 1787-88 would have concluded that the Constitution would leave such an important power unvested. Certainly no one commented on the omission. It also seems improbable that a contemporary would have accepted the alternative that the power was lodged with the executive. Contemporaries spoke and wrote in narrow terms of the President's sole martial role as Commander in Chief.[100]

(4) *As used in the Constitution, "declare" had a broader meaning than it did in the treatises and international practice. It meant*

95. *See generally* works cited in note 36 *supra.*

96. Treaty of Amity and Commerce with Prussia, Sept. 10, 1785. 8 Stat. 84, 94-96, T.S. No. 292, 2 TREATIES AND OTHER INTERNATIONAL ACTS, *supra* note 85, at 162, 178-79. This provision followed the instructions of Congress. *See* 26 JOURNALS OF THE CONTINENTAL CONGRESS 1774-1789, at 358-59 (1928).

97. *See* T. PAINE, COMMON SENSE AND OTHER POLITICAL WRITINGS 22-23 (Liberal Arts Press ed. 1953). *See generally* the works cited in note 36 *supra.*

98. *See, e.g.,* T. BAILEY, A DIPLOMATIC HISTORY OF THE AMERICAN PEOPLE 52-65 (8th ed. 1969).

99. *See, e.g.,* R. HOFSTADTER, THE AMERICAN POLITICAL TRADITION AND THE MEN WHO MADE IT 3-17 (1954).

100. *See* pp. 12-13, 19-21 *supra.*

"commence." This interpretation, unlike the first three possibilities I have discussed, would probably have seemed plausible to someone in 1787-88, because contemporary statements suggested that the power of the new Congress to commence war would be at least as broad as that of the Confederation Congress.[101] This deviation from international usage would have seemed proper, as well, since the Constitution involved domestic arrangements.

(5) *Whatever the scope of the term "declare" as used in the Constitution, any war-commencing power not covered by it was vested in Congress by virtue of that body's control of reprisals.* This possibility, which is at once distinguishable from, yet compatible with interpretation (4), would also have seemed plausible because the treatises closely assimilated imperfect war and reprisals.

The last interpretation suggests the need for more detailed consideration of how an observer, acquainted with the treatises and the relevant history, would have interpreted the Constitution's grant to Congress of power to issue letters of marque and reprisal. Here, too, there are several possibilities:

(1) *The phrase "letters of marque and reprisal" was used in the Constitution in a technical sense and was intended to give Congress authority to grant such letters to private individuals in both peace and war.* This interpretation, however, ignores the fact that special letters of marque and reprisal, for the peacetime satisfaction of private claims, had fallen into disuse and general reprisals had become the rule. As a necessary and proper concomitant of the powers to commence and conduct war, letters could be issued in war whether or not the Constitution mentioned them. In fact, if granted during a declared war as a military measure, there could be little objection to their being granted by the President, who, as Commander in Chief, was charged with the conduct of war. Under this interpretation, then, the phrase would have been archaic and redundant in intent.

(2) *The power was merely a careless and meaningless carryover from the Articles of Confederation.*[102] A contemporary would have noted, though, that the Convention had given new attention

101. *See* pp. 18-19 *supra.*

102. Of course, what we know about the history of the clause in the Philadelphia Convention (*see* pp. 12-13 *supra*) was generally unknown in 1787-88.

to the problem of letters of marque and reprisal. The states, which could issue such letters in time of war under the Articles, were altogether prohibited from issuing them under the Constitution.[103] Thus it would have seemed unlikely that the retention of the power was meaningless.

(3) *The phrase most importantly conferred on Congress power over general reprisals outside the context of declared war.* While the wording in question admittedly spoke broadly of granting "letters of marque and reprisal," issuance of the *special* variety had passed out of fashion in peace time. The clause thus could easily have been interpreted as serving as a kind of shorthand for vesting in Congress the power of *general* reprisal outside the context of declared war.[104] For someone in the late 1780's, this interpretation, far more than the first two, would have given the phrase meaning and would have been consistent with history and the treatises. Once accepted, this interpretation in turn would have given increased plausibility to the view that Congress possessed whatever war-commencing power was not covered by the phrase "to declare war."

In short, while one cannot pretend that the matter is beyond all doubt, it seems plain that knowledge of the theory and practice of war and reprisal would have helped convince a late-eighteenth century American that the Constitution vested Congress with control over the commencement of war, whether declared or undeclared.

103. ARTS. OF CONFED. art. VI; U.S. CONST. art. I, § 10.

104. The making [of] a reprisal on a nation is a very serious thing. Remonstrance and refusal of satisfaction ought to precede; and when reprisal follows, it is considered an act of war, and never failed to produce it in the case of a nation able to make war; besides, if the case were important and ripe for that step, Congress must be called upon to make it; *the right of reprisal being expressly lodged with them by the Constitution, and not with the Executive.*

Opinion of Thomas Jefferson, Sec'y of State, May 16, 1793, *quoted in* 7 J. MOORE, INT'L LAW DIGEST 123 (1906) (emphasis added). On the matter of the wording regarding reprisals serving as shorthand, *cf.* L. LEVY, ORIGINS OF THE FIFTH AMENDMENT 430 (1968): "[C]onstitution-makers, in that day at least, did not regard themselves as framers of detailed codes. To them the statement of a bare principle was sufficient . . . ." The same point might also be made in connection with the equation of "to declare war" and "to commence war."

D. *English Influences*

If, as the preceding discussion indicates, Americans in 1787-88 saw Congress as having the dominant voice in the commencement of war, they were breaking with English constitutional theory under which war-making was a Crown prerogative.[105] This is not surprising, for the scope of executive power in America narrowed considerably after independence. In the late 1780's, of course, the trend toward weaker executives was reversed. The reversal, however, resulted from domestic considerations–stronger and more independent leadership seemed necessary to insure liberty and stability *within* the country–and it had little or no connection with external problems of war-making.[106] Departure from English theory concerning the initiation of war was therefore consistent with broader developments in post-independence America.

Although the English theoretical model lost favor in America, prior English experience and thought were still quite relevant to the evolving American view of war-making. Even in England itself, practice did not coincide with theory. In the seventeenth century, Sir Matthew Hale had written: "The power of making war or peace . . . in England is lodged singly in the King, *tho it ever succeeds best when done by parliamentary advice.*"[107] In 1775 another student of the English constitution, Jean de Lolme, commented that the King

> has the prerogative of commanding armies, and equipping fleets–but without the concurrence of his Parliament he cannot maintain them . . . . He can declare war–but without his Parliament it is impossible for him to carry it on. In a word, the Royal prerogative, destitute as it is of the power of imposing taxes, is like a vast body, which cannot of itself accomplish its motions . . . .[108]

Hale and de Lolme, of course, were not directly considering the theoretical locus of the war-making power in England. They as-

105. 1 W. Blackstone, Commentaries *249-50.
106. *See* G. Wood, The Creation of the American Republic 1776-1787 *passim* and especially at 393-564 (1969).
107. 1 Hale, *supra* note 90, at 159 (emphasis added).
108. J. De Lolme, The Constitution of England 48 (1775). *See generally id.* at 47-62.

serted only that parliamentary control over the war-supporting function provided practical checks on the King's prerogative. On the basis of such analyses, however, the English system did not present a polar contrast to the evolving American notions of congressional control over war-making. Rather, it contained elements which, on the one hand, were imitated in the American Constitution with its provisions for a clear legislative monopoly of the war-supporting function, and, on the other, suggested *tendencies* toward legislative control of the war-making function itself.[109]

Other English sources which Americans probably noted would have led them to similar conclusions about the desirability of a diminished executive role in war-making. The English commonwealthmen of the seventeenth and eighteenth centuries, whose views were especially appealing in America, admittedly did not emphasize the problem of war-making as such. Their major concern regarding military matters was that standing armies posed a domestic threat. Their outlook nonetheless generally favored a diminished executive role.[110] The same was true of the broader group of Whig writers who proved so popular in revolutionary America.[111] In 1774, James Burgh went so far as to imply that Parliament should have a substantial and independent voice not only in the supporting of war, but also in its commencement, con-

109. Of course, the development of cabinet government in England blocked the growth of an independent parliamentary check on the executive. The cabinet, with or without the Crown's connivance, came both to exercise executive functions and to manage Parliament. But to the extent that they were aware of these tendencies toward cabinet government, eighteenth century Americans deprecated them as manifestations of the corruption—that is, the improper use of influence—which they thought was so generally menacing to a proper balance in government and hence to liberty. *See* B. Bailyn, The Origins of American Politics 14-58 (1968); M. Thomson, A Constitutional History of England 1642-1801, at 353-84 (1938).

110. *See* C. Robbins, The Eighteenth Century Commonwealthman (1959). But at least one of the commonwealthmen, Henry Neville (1620-1694), was explicit in his desire to "take from the king the power of making war and peace . . . ." *Id.* at 39. Far from a coherent group, the English Commonwealthmen or Real Whigs were intellectual descendants of the Puritan revolutionaries of the 1640's and 1650's. They had republican and non-conformist tendencies and were enamored of natural right theories, freedom of thought, limited government, and parliamentary reform. By the end of the eighteenth century, they had merged into the broader stream of English radicalism. *See id. passim.*

111. *See* Colbourn, *supra* note 69, *passim.*

duct, and conclusion.[112] John Locke, writing some ninety years earlier, had separated what he called the federative power, which included war-making, from the ordinary executive power. He saw the same authority as properly exercising both powers, but his separation of them for purposes of analysis may have aided in establishing a basis for their later actual separation.[113] At any rate, Thomas Rutherforth was explicit in his opinion in the 1750's that war-making was a legislative function because it had to rest on the "common understanding" of the nation.[114] All considered, it is not surprising to find that at the Philadelphia Convention James Madison and James Wilson characterized war-making as properly legislative.[115]

III. Conclusion: The Original Understanding in Theory and Early Practice

Although the change from "make" to "declare" in the clause empowering Congress "to declare War" was open to several interpretations among the members of the Philadelphia Convention, there is enough evidence to allow some cautious generalization about the original understanding concerning war-making. The Confederation Congress exercised both legislative and executive functions; the new Congress would not. Nevertheless, specific remarks equating the war-making powers of the two Congresses, together with other comments about war-making being a legislative function, suggest that contemporaries thought the power of the proposed legislature to commence war would be as broad as that of the Confederation Congress. Since the old Congress held blanket power to "determine" on war, and since undeclared war was hardly unknown in fact and theory in the late eighteenth century, it therefore seems a reasonable conclusion that the new Congress'

112. *See* 1 J. Burgh, Political Disquisitions: Or An Enquiry Into Public Errors, Defects, and Abuses . . . 371 (1774). *See also id.* at 193-95, 414-45.

113. *See* J. Locke, The Second Treatise of Government 83-84 (Liberal Arts Press ed. 1952).

114. 2 Rutherforth 64-67.

115. *See* p. 11 *supra*. *See also* 1 The Works of James Wilson, *supra* note 78, at 433-34 (Wilson's 1790-91 law lectures).

power "to declare War" was not understood in a narrow technical sense but rather as meaning the power to commence war, whether declared or not. To the extent that the power was more narrowly interpreted, however, the new Congress' control over letters of marque and reprisal must have suggested to contemporaries that it would still control "imperfect"–that is, undeclared–war. Otherwise, the provision involving such letters would have seemed practically meaningless in view of the status of reprisals by the late 1780's. Taken together, then, the grants to Congress of power over the declaration of war and issuance of letters of marque and reprisal likely convinced contemporaries even further that the new Congress would have nearly complete authority over the commencement of war. Reinforcing the same conclusion is the fact that English experience, and particularly the English Whigs to whom Americans paid considerable attention, offered hints about the desirability of legislative supremacy in this area. It remains possible that the President as Commander in Chief was *tacitly* accorded the initiative to meet sudden attacks on the United States. In their public statements, however, contemporaries assigned him the restricted military role of conducting war once it had begun. In any event, the Constitution explicitly authorized the states to act in the face of surprise attack.

The consensus which existed in 1787-88 on the war-making issue did not last, and Hamilton as "Pacificus" began lodging reservations as early as 1793. It is significant, though, that his position appears not to have been an especially persuasive one even to his fellow Federalists,[116] let alone to the Republicans. In 1795, James Kent contended that in the United States "war only can be commenced by an act or resolution of Congress," indicating that he equated "declare" with "commence."[117] The undeclared hostilities with France during John Adams' administration were authorized by Congress,[118] and despite debates over the extent to which specific legislative measures provided a basis in law for action

116. That is, members of the Federalist Party, not to be confused with the lower case "f" federalists of the ratification controversy in 1787-1788.

117. KENT, *supra* note 87, at 83.

118. *See* A. DECONDE, THE QUASI WAR: THE POLITICS AND DIPLOMACY OF THE UNDECLARED WAR WITH FRANCE 1797-1801 *passim* and especially at 89-98 (1966).

against the French, the notion that authorization must come from Congress was seldom challenged.[119] At one point, when the Republicans in the House of Representatives proposed to modify and weaken legislation authorizing American naval reprisals against the French, an exasperated Federalist protested that preservation of American rights required the legislation to be passed unamended, because "the President has not the power to act in the [present] case. Congress only [can] authorize reprisals."[120] Retreating from the position he had earlier taken as "Pacificus," Alexander Hamilton held that the Constitution narrowly constrained the President's actions. Adams might authorize the repelling of actual attacks, but he could not make reprisals without congressional approval.[121]

In two maritime prize cases arising out of the Quasi-War, the Supreme Court of the United States evinced similar views. The seriatim opinions in *Bas v. Tingy* (1800)[122] stressed that whether hostilities were declared or undeclared, they still constituted war—being perfect and general war in the one case, and imperfect and limited war in the other. None of the Justices explicitly stated that only Congress might wage imperfect war, but that conclusion was clearly implicit in their remarks.[123] In *Talbot v. Seeman* (1801), a case involving the salvage rights of Silas Talbot and his officers and crew, John Marshall, the newly appointed Chief

119. *See, e.g.*, THE DEBATES AND PROCEEDINGS IN THE CONGRESS OF THE UNITED STATES, 5th Cong., 2d Sess. cols. 1440-1522, 1783, 1798-1812, 1815-35 (1798).

120. *Id.*, col. 1828 (Representative James A. Bayard).

121. 1 NAVAL DOCUMENTS RELATED TO THE QUASI-WAR BETWEEN THE UNITED STATES AND FRANCE 75-76 (1935) (letter from Hamilton to Sec'y of War J. McHenry, May 17, 1798). *See also id.* at 78 (letter from Sec'y of War J. McHenry to Captain R. Dale, USN, May 22, 1798). On the general conclusions contained in this paragraph, *see* Russell, *supra* note 6, at 65-100 (for the Federalist period 1789-1801) and 102-46 (for the Republican period, 1801-1815).

122. 4 U.S. (4 Dall.) 37 (1800).

123. *See, e.g.*:

> Congress is empowered to declare a general war, or congress may wage a limited war; limited in place, in objects, and in time. If a general war is declared, its extent and operations are only restricted and regulated by the *jus belli*, forming a branch of the law of nations; but if a partial [war] is waged, its extent and operation depend on our municipal laws [as passed by Congress].

Id. at 43 (Chase, J.).

Justice, forthrightly stated: "The whole powers of war being, by the Constitution of the United States, vested in congress, the acts of that body alone be resorted to as our guides in the enquiry." He further argued that such powers included the authorization of limited hostilities, which he, too, obviously regarded as war, for he explicitly referred to the recent conflict with France as "war."[124]

Evidence from the years immediately following ratification of the Constitution thus corroborates the conclusion that Americans originally understood Congress to have at least a coordinate, and probably the dominant, role in initiating all but the most obviously defensive wars, whether declared or not. Since that time, and especially during the twentieth century, the presidential role in war-making has nevertheless become dominant. Somewhat in the manner of Teddy Roosevelt and the Panama Canal, while Congress and others have debated, Presidents have acted. How long this situation will continue is difficult to predict, though the constitutionality of recent practice is an intriguing question which promises to attract further national attention.[125] Whether the original understanding properly concludes the issue undeniably involves questions quite different from those I have here discussed. Still, paying it some heed is surely consonant with a devotion to constitutionalism.

124. Talbot v. Seeman, 5 U.S. (1 Cranch) 1, 28, 32 (1801). On the conflict's status as war, *see* 1 OP. ATT'Y GEN. 84 (1798) (C. Lee to Sec'y of State).

125. On April 13, 1972, the Senate passed a bill to regulate undeclared war. *See* S. 2956, 92d Cong., 1st Sess. (1971) ("A bill to make rules governing the use of the Armed Forces of the United States in the absence of a declaration of war by the Congress"). N.Y. Times, Apr. 14, 1972, at 1, col. 6 (city ed.). Though the bill passed 68-16, resistance from both the House and the Executive is likely. *See* Bickel, *The Need for a War-Powers Bill,* NEW REPUBLIC, Jan. 22, 1972, at 17. For earlier bills and resolutions of this sort, see, *e.g.,* S. 731, 92d Cong., 1st Sess. (1971); S. J. Res. 95, 92d Cong., 1st Sess. (1971); and nineteen House bills and resolutions reprinted in *1970 Hearings on Congress, the President, and the War Powers, supra* note 4, at 435-76.

2

COMPULSORY MILITARY SERVICE UNDER THE CONSTITUTION:

The Original Understanding

Did the Constitution's founders intend to permit compulsory military service? In 1918 the United States Supreme Court answered with a ringing yes, arguing that "as the mind cannot conceive an army without the men to compose it, on the face of the Constitution the objection that it does not give the power to provide for such men would seem to be too frivolous for further notice." Fifty years later the Court saw no reason to alter this view.[1] In recent decades others have found the question less frivolous: constitutional debate, including discussion of the original understanding, occurred when Selective Service was established in 1940, when Universal Military Training was proposed at the end of World War II, and when conscription came under attack in the final years of the Indochina War.[2] Now the draft has ended, but the

1. *Selective Draft Law Cases,* 245 U. S. 366, 377-378 (1918); *United States* v. *O'Brien,* 391 U.S. 367, 377 (1968).

2. See, for example, Claude B. Mickelwait, "Legal Basis for Conscription," *American Bar Association Journal,* XXVI (1940), 701-705, Harrop A. Freeman, "The Constitutionality of Peacetime Conscription," *Virginia Law Review,* XXXI (1944), 40-82; W. Randolph Montgomery, "The Relation of the Militia Clause to the Constitutionality of Peacetime Compulsory Universal Military Training," *ibid.,* 628-666; Leon Friedman, "Conscription and the Constitution: The Original Understanding," *Michigan Law Review,* LXVII (1969), 1493-1552; Harrop A. Freeman, "The Constitutionality of Direct Federal Military Conscription," *Indiana Law Journal,* XLVI (1971), 333-346; Anthony A. D'Amato and Robert M.

questions whether the Americans who framed and ratified the Constitution intended compulsory military service to exist under it, and, if so, in what form, remain historically interesting. The difficulty in answering them lies in avoiding anachronism and even more in interpreting some large silences on the part of the contestants of 1787-1788.

The evidence on a preliminary point is conclusive. Late eighteenth-century Americans agreed, at least in theory, that government could lawfully require citizens to perform *some* manner of military service, sometimes with allowance for alternative service or monetary payments.[3] The Massachusetts Bill of Rights of 1780 expressed the basic sentiment: "Each individual of the society has a right to be protected by it in the enjoyment of his life, liberty, and property, according to standing laws. He is obliged, consequently, to contribute his share to the expense of the protection;

O'Neil, *The Judiciary and Vietnam* (New York, 1972), 89-97; and Michael J. Malbin, "Conscription, the Constitution, and the Framers: An Historical Analysis," *Fordham Law Review,* XL (1972), 805-826. For attempts to raise the constitutional and original understanding issues in court during the Indochina conflict see, for example, *Massachusetts* v. *Laird,* 400 U. S. 886 (1970), and *United States* v. *Diaz,* 427 F.2d 636, 639 (1st Cir. 1970). For pre-World War II debates over compulsory military service, which often raised constitutional issues, see Jack F. Leach, *Conscription in the United States: Historical Background* (Rutland, Vt., 1952). A useful overview is Joseph C. Duggan, *Legislative and Statutory Development of the Federal Concept of Conscription for Military Service* (Washington, D. C., 1946). These works illustrate the debate on the question—a debate from which historians have largely stood aside. They further contain much useful information and are sometimes of high quality. See especially the articles by Friedman and Freeman (taking an antidraft position) and Malbin (taking a prodraft position), which, however, still do not always address the crucial issues. They also misinterpret some evidence and take too little notice of qualifications which the evidence requires. Nevertheless, I have profited from them and freely acknowledge my debt.

3. For a compilation of the military laws of the colonies and states see Selective Service System, *Backgrounds of Selective Service,* Special Monograph No. 1, II (Washington, D. C., 1947), Pts. 2-14, *passim.* State constitutional provisions are in Benjamin P. Poore, comp., *The Federal and State Constitutions, Colonial Charters and Other Organic Laws of the United States* (Washington, D. C., 1878). These provisions ranged from highly explicit statements of obligation (as in Massachusetts, New Hampshire, New York, and Pennsylvania) to mere passing references to the militia (as in New Jersey and South Carolina), which must be read in light of the militia laws of the states in question. See also studies cited in n. 5 below.

to give his personal service, or an equivalent, when necessary."[4] In law all able-bodied white males of appropriate age—sixteen to sixty, for example—were members of the colonial or state militias. In practice before the Revolutionary War some free men avoided enrollment in the militia, the militia system itself decayed over the colonial period, and actual military forces were raised perhaps more often by enlisting volunteers than by drafting from the militia.[5] But this at most shows that the colonies did not always compel in fact the service due by law.

The legal permissibility of some form of compulsory military service in late eighteenth-century America provides, however, only a general context for reconstructing the founders' *constitutional* understanding in 1787-1788. The problem is to determine what *specific* powers relating to military service they intended to assign to *specific* governments. A largely neglected debate in the House of Representatives during the Quasi-War with France hints at an approach.

In May 1798 the House was considering a bill supported by the Adams administration to establish a provisional army; the section of immediate interest would have allowed the president to accept units of volunteers into the army if he determined that they would be useful. Objecting to this, Republicans pointed to the local ties of the projected units, to the fact that service in them could be substituted for militia service, and to the likelihood that they would draw manpower from the militia. Hence, it was argued, the proposed units should be regarded not as army troops but as militia, subject to the existing constitutional limitations on the militia.[6]

4. Poore, comp., *Federal and State Constitutions,* I, 958.

5. Frederick P. Todd, "Our National Guard: An Introduction to Its History," *Military Affairs,* V (1941), 73-76; Louis Morton, "The Origins of American Military Policy," *ibid.,* XXII (1958), 75-82; John W. Shy, "A New Look at Colonial Militia," *William and Mary Quarterly,* 3d Ser., XX (1963), 175-185; Douglas Edward Leach, *Arms for Empire: A Military History of the British Colonies in North America, 1607-1763* (New York, 1973).

6. [*Annals of Congress*] *The Debates and Proceedings in the Congress of the United States* (Washington, D. C., 1834-1856), VIII, cols. 1703-1706, 1725-1767, hereafter cited as *Annals of Congress.* There are two editions of the *Annals,* which are paginated differently. All citations are to the edition that has the running head "History of Congress" on *each* page. The Republicans also lodged other constitutional objections—for example, that the provision delegated legislative power to the president.

The Federalists were not persuaded. Samuel W. Dana of Connecticut declared that he "could not consider these volunteer corps a part of the militia," because "if there be any characteristic marks by which to distinguish a militia from regulars, it is that the former are compelled to serve, and the latter enter voluntarily into service."[7] Other Federalists agreed with Dana.[8] In all likelihood most contemporaries would have added further distinguishing traits, such as the soldier's low character and the militiaman's passion for liberty. It seems significant, though, that the more crucial part of Dana's distinction, in terms of clarifying what is historically least clear, had already received recognition from a Republican, William C. C. Claiborne of Tennessee, who stigmatized armies as comprising those "who fight *only for pay.*"[9] And in immediate response to Dana, Republican Joseph McDowell of North Carolina, while still opposing the volunteer units, seems implicitly to have accepted the distinction.[10] Another Republican, Albert Gallatin of Pennsylvania, gave it explicit sanction, although he, too, opposed the provision under discussion. Because the proposed force combined army and militia characteristics, it fell, he claimed, "about half way between a regular standing army and a militia" and thus was outside the only two descriptions of armed force recognized in the Constitution. Among the army characteristics Gallatin attributed to the troops was the fact that "they are to be enlisted (not properly so by name, but in a way equally effectual, *by giving their consent to serve*)."[11]

7. *Ibid.,* 1704.

8. *Ibid.,* 1704-1705 (Robert G. Harper, S. C.), 1733 (Harrison Gray Otis, Mass.).

9. *Ibid.,* 1651. Claiborne did refer to army troops as "slaves," and he described militiamen as animated by love of country, which superficially suggests a view opposite to the one attributed to him here. A careful reading reveals, however, that he used "slaves" to indicate that soldiers exercised no independent judgment as to the merits of the cause for which they fought so long as they were paid, which entails the conclusion that army service was legally voluntary. While Claiborne did not comment on the legal basis of militia service, his remarks on militiamen were compatible with the idea of compulsory militia duty as an obligation of citizenship.

10. *Ibid.,* 1704-1705. Although the record is not entirely clear, McDowell apparently argued that it was not at all certain that the men in question would actually be enlisted into federal service. Hence the enlistment required by Dana's scheme as a characteristic of army service would probably be missing.

11. *Ibid.,* 1725-1726, italics mine.

Coming but a decade after the ratification of the Constitution, this limited agreement on one aspect of the differences between armies and militia is highly suggestive, particularly in view of the speakers' disagreement otherwise over the legislation at issue. It leads, with a minimum of anachronism, to two more refined questions: did Americans ten years earlier, in 1787-1788, similarly understand service in "armies," as the term is used in the Constitution, to be legally voluntary; and did they regard militia duty as the sole legal basis of compulsory military service within their new frame of government? These questions provide fruitful guides for reconstructing the framers' and ratifiers' often tacit original understanding.[12]

The Federal Convention proposed a constitution which, compared with the Articles of Confederation, markedly increased federal power to raise armies and to call forth, discipline, and govern the militia of the states. The earlier instrument had admonished "every State" to "keep up a well-regulated and disciplined militia, sufficiently armed and accoutered," and had authorized the Congress "to agree upon the number of land forces, and to make requisitions from each State for its quota." It made no provision for the central government to call out the state militias, and it left to the states the execution of any requisitions for land forces. Although the Convention at Philadelphia initially devoted little attention to the problems of armies and the militia, the Committee of Detail included army and militia clauses in its draft constitution. This gave Congress power "to raise armies"; the Convention added the restriction that "no appropriation of money to that use shall be for a longer term than two years." The committee's report also authorized Congress "to call forth the aid of the militia, in order to execute the laws of the Union, enforce treaties, suppress insurrections, and repel invasions." The Convention struck " 'enforce treaties' as being superfluous since treaties were to be 'laws,' " and gave to Congress the additional power "to provide for organizing, arming, and disciplining the militia, and for governing such

12. It should be noted that the views expressed in the Philadelphia Convention do not necessarily indicate the original understanding. They may be highly suggestive, but the state debates are more conclusive legally. In any case, a still wider range of materials is usually necessary to illuminate how Americans of the day probably understood the various clauses of the Constitution.

part of them as may be employed in the service of the United States, reserving to the States respectively, the appointment of the officers, and the authority of training the militia according to the discipline prescribed by Congress." The army and militia clauses passed through the Committee of Style and into the finished Constitution without substantial change.[13]

These clauses attracted enough attention in the Convention and the ratification debates to allow one to discern, with reasonable assurance of accuracy, what it was about armies and militia and the clauses relating to them that most concerned contemporaries. Defenders of the Constitution held not only that congressional authority to raise armies was vital to national security, but that it must exist in peace as well as war. An army in being might forestall attack; waiting to raise an army until an attack occurred would be hazardous at the least. The militia would be useful against foreign foes and in supporting the federal government in its domestic functions, but it was not adequate by itself, particularly with reference to external threats. However, to the extent that the militia was well regulated and uniform as provided by the Constitution, it would decrease the need for a standing army. Indeed, no army could endanger liberty in the face of the effective militia that the Constitution would create. Countering these points, the Antifederalists not only minimized the external dangers to American security, but portrayed the Constitution's allowance of a peacetime standing army as an opening for tyranny. Drawing different lessons from history than did the Federalists, they regarded the militia as a reliable line of defense in peacetime. They feared, however, that the militia clauses would allow the federal government to make militia service so distasteful as to turn the citizenry against the militia and thus destroy it as a bulwark of liberty. They were also concerned lest these clauses make the militia itself a threat to political freedom and perhaps to personal rights.[14]

13. Articles of Confederation, Arts. VI, IX; Max Farrand, ed., *The Records of the Federal Convention of 1787,* rev. ed. (New Haven, Conn., 1937 [orig. publ. 1911]), II, 182, 384-390, 508-509, 595; Constitution of the United States, Art. I, Sec. 8. For a detailed history of these clauses in the Convention see page references in Farrand, ed., *Records,* IV, 114.

14. For the most important statements see Farrand, ed., *Records,* II, 326-333,

That these issues were disputed is understandable. Remarkably similar debates about standing armies had aroused Englishmen during the seventeenth century and had generated a substantial literature on which eighteenth-century Americans drew heavily. The American colonists also had absorbed anti-army arguments from more recent English radical pamphleteers. When controversy with the mother country grew hot in the 1760s and 1770s, and more English troops moved into the colonies, Americans detailed the threat standing armies posed to liberty. At the same time they praised a strong militia, animated by the spirit of liberty, as a defense against tyranny. After the outbreak of hostilities, fear of armies and faith in the militia figured in Congress's failure to support fully the Continental army.[15] In sum, the dispute over the Constitution's military clauses reiterated an old theme.

Because the debate took the direction it did, it does not directly reveal as much as one would wish about views in 1787-1788 on the status of compulsory service under the Constitution. Still, the arguments of the Federalists, who, after all, won the contest, *seem*

509; Jonathan Elliot, ed., *The Debates in the Several State Conventions on the Adoption of the Federal Constitution . . .*, 2d ed. (New York, 1888), III; Jacob E. Cooke, ed., *The Federalist* (Middletown, Conn., 1961), esp. Nos. 8, 23-29, 41; *Letters from the Federal Farmer to the Republican* (New York, 1787), 23-25; and Luther Martin, "The Genuine Information, Delivered to the Legislature of the State of Maryland . . . ," 1787, in Farrand, ed., *Records,* III, 207-209. All subsequent references to *The Federalist* are to Cooke's edition; where no pages are cited, the reference is to the cited paper in general. Recent studies of the 1787-1788 army debate include Bernard Donahoe and Marshall Smelser, "The Congressional Power to Raise Armies: The Constitutional and Ratifying Conventions, 1787-1788," *Review of Politics,* XXXIII (1971), 202-211; Richard H. Kohn, "The Federalists and the Army: Politics and the Birth of the Military Establishment, 1783-1795" (Ph.D. diss., University of Wisconsin, 1968), 183-219; and Arthur A. Ekirch, Jr., *The Civilian and the Military* (New York, 1956), 24-31. These, however, do not explore the issue of compulsory service.

15. Lois G. Schwoerer, *"No Standing Armies!": The Antiarmy Ideology in Seventeenth-Century England* (Baltimore, 1974), esp. Chap. 9; Bernard Bailyn, *The Ideological Origins of the American Revolution* (Cambridge, Mass., 1967), 22-54, 112-119; John Shy, *Toward Lexington: The Role of the British Army in the Coming of the American Revolution* (Princeton, N. J., 1965), 376-398; John C. Miller, *Origins of the American Revolution* (Boston, 1943), 431-433, 440; Kohn, "The Federalists and the Army," 1-19; Don Higginbotham, *The War of American Independence: Military Attitudes, Policies, and Practice, 1763-1789* (New York, 1971), esp. 205-213.

to cast doubt on the position that Dana advanced in 1798. One of the major reasons for the Constitution, claimed the Federalists, was the Confederation's inability to raise armies; surely, therefore, the Federalists did not intend to deny this power to the new government.[16] Nor did they incline toward a narrow view of what might be necessary in this regard: they decried limiting the means available to the federal government when future contingencies were so unpredictable.[17] On this point the Federalist spokesmen and their audiences must have recalled the need for a draft during the War for Independence. In fact, Alexander Hamilton implied that the Constitution gave the Union the power to "levy" troops.[18] In asking that recognition be given to the right of conscientious objection, three ratifying conventions similarly revealed their assumption that under the Constitution the Congress would have power to draft men into the army. Displaying the same premise was a recommendation "that no person shall be compelled to do military duty otherwise than by voluntary enlistment, except in cases of general invasion."[19] Antifederalist comments also reflected the opinion that there would be few limits to the proposed government's power to raise armies.[20] As for the militia, neither Federalists nor Antifederalists explicitly stated that it was the sole form of compulsory service, and neither side suggested that the militia's constitutional status restricted army service.

Whether they approved or not, few politically aware Americans in 1787-1788 could have doubted that the Constitution's "plain words" conveyed broader federal power over armies and state militias than did the Articles of Confederation. This, however, does not mean that they concluded that the clause giving Congress the power "to raise and support armies" carried authority to com-

16. Farrand, ed., *Records,* I, 19-25; *The Federalist,* No. 22, 137-138; Elliot, ed., *Debates,* II, 520-521, III, 91, IV, 96-99; Frederick W. Marks III, *Independence on Trial: Foreign Affairs and the Making of the Constitution* (Baton Rouge, La., 1973), 142-206.

17. *The Federalist,* No. 23, No. 41, 270.

18. *Ibid.,* No. 23, 148.

19. Elliot, ed., *Debates,* I, 335, 336 (R. I.), III, 659 (Va.), IV, 244 (N. C.).

20. See, for example, *Letters from the Federal Farmer,* 24, and John Bach McMaster and Frederick D. Stone, eds., *Pennsylvania and the Federal Constitution, 1787-1788* (Lancaster, Pa., 1888), 154-155.

pel service. They ignored the issue in open debate;[21] had they discussed it, they probably would have rejected the conclusion. But they indicated awareness that the militia clause, unlike the army clause, did offer the new government a limited means, which they seem to have considered its sole means, of compelling service. In short, it is likely that Dana's distinctions in 1798 correctly represented the common view of a decade earlier. Evidence warranting this conclusion comes from remarks about armies found in the 1787-1788 debates and other sources from the period; comparisons therein drawn between militia and armies; implications flowing from Revolutionary, Confederation, and foreign models; assumptions revealed by several proposed amendments; and a contextual examination of certain open-ended Federalist comments.

By 1787-1788 the Confederation's limitations in raising military forces had become reasonably clear. Elbridge Gerry, who refused to sign the Constitution and opposed its ratification, observed in Philadelphia that "the existing Congs. is so constructed that it cannot of itself maintain an army." At a descriptive level the Federalists agreed. It was not that the Confederation lacked basic authority, they claimed; the problem was rather one of implementation. The quota system, under which Congress called upon each state to furnish a specified number of troops, had proved inefficient and inequitable during the war. The states had varied in their compliance with congressional requests. In raising troops, moreover, they had competed with each other and with Congress in offering bounties to enlistees; and Congress, which depended on the states for financial resources, could not outbid them in this regard.[22] Difficulties had continued into the postwar

21. Edmund Randolph and Alexander Hamilton made passing references to a draft. See Farrand, ed., *Records,* I, 19, 25, and *The Federalist,* No. 22, 138. These produced no recorded discussion of the subject, insofar as the surviving evidence indicates; the two men were arguing that other resources were needed so the new government could *avoid* a draft; and in any case their references were to the Revolutionary War draft, which was not an army draft.

22. Farrand, ed., *Records,* II, 329; *The Federalist,* Nos. 22-23, No. 30, 189, No. 40, 262-263; Elliot, ed., *Debates,* II, 213-214, 468; James Iredell, *Answers to Mr. Mason's objections to the new Constitution* (Newbern, N. C., 1788), reprinted in Paul L. Ford, ed., *Pamphlets on the Constitution of the United*

period. These were not so specifically mentioned by the Federalists in 1787-1788 but were surely on their minds. Contending that the Articles gave Congress no express power even to *requisition* a peacetime standing army, the New England states had blocked an attempt to raise regular federal troops in June 1784 and thereby forced Congress to resort to a *recommendation* for state militia.[23] To this and subsequent congressional calls for forces the states were slow to respond, and the troops so raised fell under joint congressional and state control. At one point during Shays's Rebellion, Massachusetts went so far as to countermand orders from Secretary at War Henry Knox to federal forces within the state.[24]

The obvious remedy for such limitations was simply, but importantly, to eliminate the states as federal recruiting and revenue agents. This was what Hamilton proposed in *The Federalist,* when he argued for "discard[ing] the fallacious scheme of quotas and requisitions" and held "that the Union ought to be invested with full power to levy troops; . . . and to raise revenues, which will be required for the formation and support of an army . . . , in the customary and ordinary modes practiced in other governments."[25] The most important of these requirements was revenue. Elsewhere in *The Federalist* James Madison implicitly conceded that with adequate revenue even the Confederation could raise a substantial force.[26] Reiterating Edmund Randolph's opinion at the beginning of the Philadelphia Convention, Hamilton called revenue "the essential engine by which the means of answering the

---

States . . . 1787-1788 (Brooklyn, N. Y., 1892), 363; Oliver Ellsworth, "Letters of 'A Landholder,'" in Paul L. Ford, ed., *Essays on the Constitution of the United States . . . 1787-1788* (Brooklyn, N. Y., 1892), 157. These comments, especially the more allusive and general ones, should be read in light of the actual weakness revealed by the wartime requisition and quota system, discussed by Higginbotham, *War of American Independence,* 288-309, 389-412.

23. *Journals of the Continental Congress, 1775-1789* (Washington, D. C., 1904-1937), XXVII, 433-434, 538-539. See also Edmund C. Burnett, ed., *Letters of Members of the Continental Congress* (Washington, D. C., 1921-1938), VII, 542-543, 546-547, 552, 588, 594-595, 604-605.

24. Harry M. Ward, *The Department of War, 1781-1795* (Pittsburgh, Pa., 1962), 42-43, 56-57, 75-81. For congressional resolutions to raise forces see *Jour. Cont. Cong.,* XXVIII, 223-224, 240, 247-248, XXXI, 891-893, XXXIII, 603.

25. *The Federalist,* No. 23, 148-149.

26. *Ibid.,* No. 38, 248-249.

national exigencies must be procured."[27] Furthermore, in approving the "power to *levy* troops" he did not endorse a draft.[28] Antifederalists, too, thought money the key.[29] The debate on the two-year limitation on army appropriations showed that both sides agreed on the importance of revenue,[30] as did the dispute over whether the Constitution adequately separated the powers of purse and sword.[31]

While it may appear strange that men who had recently seen Congress forced to resort to a draft would focus so exclusively on revenue as the "essential engine" of defense, the emphasis is understandable. The Revolutionary draft had occurred when the central government lacked an adequate revenue.[32] Also, contemporaries most likely agreed with the "Federal Farmer" that, given adequate funding, "an army is a very agreeable place of employment for the young gentlemen of many families."[33] Hamilton men-

27. Farrand, ed., *Records,* I, 19, 25; *The Federalist,* No. 31, 196. For similar remarks on revenue see *ibid.,* No. 30, 191-192, No. 41, 276, and Elliot, ed., *Debates,* II, 350-351, 361-370.

28. Although to the modern ear "levy" suggests compulsion, the context of its use by Hamilton in *The Federalist,* No. 23, indicates a meaning equivalent to that of "raise," and the dictionaries of the period indicate the same meaning. See, for example, Samuel Johnson, *Dictionary of the English Language,* 2d ed. (London, 1755), II, n.p., which defines "levy" as "to raise, to bring together men." Moreover, in *The Federalist,* No. 22, 138, Hamilton spoke of "short inlistments" as a form of "levies." See also *Laws Enacted in the First Sitting of the* [*Pennsylvania*] *Twelfth General Assembly . . . 1787* (Philadelphia, 1787), 402, Sec. 2, for an example of the use of "levy" in a recruiting act restricted to voluntary enlistment.

29. See, for example, Elliot, ed., *Debates,* II, 333; *Letters from the Federal Farmer,* 24; [Samuel Bryan?], "Letters of Centinel," in McMaster and Stone, eds., *Pennsylvania and the Federal Constitution,* 587-588.

30. *The Federalist,* Nos. 24, 26, 41; Elliot, ed., *Debates,* II, 536-537, IV, 97-98; letter by "Plain Truth," in McMaster and Stone, eds., *Pennsylvania and the Federal Constitution,* 191; "Letters of Centinel," *ibid.,* 585; Farrand, ed., *Records,* II, 509.

31. Elliot, ed., *Debates,* I, 54-107, *passim,* II, 57, 61-67, 132, 348-351, 375, III, 76-79, 128, 393-394, 410, 600, IV, 169, 214, 220.

32. For Federalist recognition of this see *The Federalist,* No. 22, 138, No. 23, 148; Farrand, ed., *Records,* I, 19, 25. See also *The Federalist,* No. 30, for related comments.

33. *Letters from the Federal Farmer,* 24. In editing these letters Paul Ford at this point incorrectly inserted a "not," making the sentence read: "An army is not a very agreeable place of employment." Ford, ed., *Pamphlets,* 304-305. His error has been repeated elsewhere. See, for example, Cecelia M. Kenyon, ed., *The Antifederalists* (Indianapolis, 1966), 227.

tioned "the young and ardent." George Washington had already written that men between eighteen and twenty-five possessed "a natural fondness for Military parade (which passion is almost ever prevalent at that period of life)."[34] Perhaps these gentlemen recalled what scattered evidence suggests was the situation before the War for Independence: despite their aversion to standing armies, which has been better remembered (and perhaps exaggerated) by later generations, Americans had enlisted in and indeed had been fascinated by the British army.[35] But, as Oliver Ellsworth asked, "Who will go to war and pay the charges of his own warfare?"[36]

What memories of the Revolutionary War should have produced, if Americans understood the armies clause as carrying the power to compel service, was comment on this feature of the clause. The Federalists might have commended it as an advantage to be gained from the new Constitution. For not only had the states performed poorly in meeting their wartime troop quotas, but they also had been lax in drafting men to fill them. However, the Federalists were silent. Aware of how the draft evoked unpleasant memories (Hamilton referred to "those oppressive expedients for raising men"),[37] they may, of course, have considered it politically unwise to dwell on this new federal power. But if many contemporaries agreed that the armies clause conveyed the power, or were even open to argument on the issue, then presumably these same unpleasant memories would have given effectiveness to an Antifederalist attack on this part of the Constitution. Yet the Antifederalists were similarly silent.[38]

Corroborating the conclusion that the disputants did not tacitly

34. *The Federalist,* No. 29, 183; Washington, "Sentiments on a Peace Establishment" (1783), in John C. Fitzpatrick, ed., *The Writings of George Washington . . . 1745-1799* (Washington, D. C., 1931-1944), XXVI, 389. Both men were discussing the formation of a select corps of militia, but the view of each would be just as applicable to army service.

35. Shy, *Toward Lexington,* 143, 145, 354, 360-362, 384-385, 392-394. Note that the "drafting" Shy mentions at p. 360 is drafting from one army regiment into another and not from civilian life into the army.

36. "The Letters of a Landholder," in Ford, ed., *Essays,* 157. James Wilson made the same point. See Elliot, ed., *Debates,* II, 522.

37. *The Federalist,* No. 22, 138.

38. In fact, the Revolutionary draft was a particular kind of draft, which, among other things, probably explains why memory of it did not produce debate

accept compulsory army service, and giving more insight into the importance they assigned to revenue, is their view that "armies" were professional forces. Almost without exception in the constitutional debates, both sides used the term "army" interchangeably with "standing army."[39] The Federalists, in particular, might have deflected criticism of Congress's power to raise armies by developing a distinction between citizen armies and standing armies. They instead admitted the dangers of armies and then proceeded to argue that armies were needed anyway and that the Constitution would control the attendant dangers.[40]

Most significantly, the standing armies that so monopolized the vision of contemporaries were thought to consist of base men often held by harsh discipline but also motivated by monetary reward. In 1778 Washington had recognized that the "common, received Opinion" about armies derived from the idea of "their being Mercenaries; hirelings."[41] A popular drill manual published on the eve of the Revolution held that American militia units comprised men who thought for themselves, while "standing armies are composed of very different men. These serve only for their pay."[42] James Burgh, the English whig whose writings were widely read and favorably received in America, deprecated standing armies for similar reasons. An Antifederalist's disparaging

---

on the armies clause, but consideration of this aspect is best postponed to a later section. See text accompanying nn. 56-61 below for a fuller discussion.

39. I base this generalization on my reading of the literature cited in n. 14 above. For specific examples see *The Federalist,* Nos. 8, 24-26; *Letters from the Federal Farmer,* 23-24. For two exceptions, which, however, do not vitiate the point in the text, see n. 45 below.

40. See Donahoe and Smelser, "Power to Raise Armies," *Review of Politics,* XXXIII (1971), 204-208, for a survey of the arguments. A partial exception is in [Tench Coxe], "An Examination of the Constitution . . . ," in Ford, ed., *Pamphlets,* 151.

41. Washington to John Banister, Apr. 21, 1778, in Fitzpatrick, ed., *Writings of George Washington,* XI, 290-291. Washington also denied that the common opinion accurately described the Continental army, which was formed of "Citizens, having all the Ties, and interests of Citizens, and in most cases, property totally unconnected with the Military Line." *Ibid.,* 291. In the main, however, Americans of the 1780s saw only two alternatives: a professional force or militia. See Kohn, "The Federalists and the Army," esp. 219-221.

42. Timothy Pickering, *An Easy Plan of Discipline for a Militia* (Salem, Mass., 1775), Pt. 1, 10. Note the close similarity between this comment and that by Claiborne in 1798, quoted above.

reference in 1787 to a standing army as "a hired military force" suggests a continued American receptivity to the theme. Still more telling is Hamilton's contrast between the militia, possibly serving by compulsion, and a regular force "in pay of the government."[43]

Although it accorded with British usage in the period, the assimilation of "armies" and "standing armies" in the constitutional debates may seem odd because in both England and America "army" also retained its older and broader meaning of simply a military force in being. In some instances Americans even used the term to describe a joint force of regulars and militia.[44] Yet whatever advantages it might have given them, the Federalists had good reason not to read the broader, less pejorative meaning into the armies clause.[45] Contemporaries could have concluded that such an interpretation made redundant the authority granted Congress to call up state militias and that it nullified the Constitution's restrictions on such calls. This, of course, would have contradicted specific Federalist reassurances that the proposed government could call out the militia only for the carefully defined purposes of executing the law, suppressing insurrections, and repelling invasions.[46]

Because the founders thought that state militia service would be

43. James Burgh, *Political Disquisitions: or, An Enquiry into Public Errors, Defects, and Abuses . . .* (London, 1774-1775), II, 343-346, 370-371, 397; "Letters of Centinel," in McMaster and Stone, eds., *Pennsylvania and the Federal Constitution,* 585; *The Federalist,* No. 24, 156-157.

44. See, for example, James A. H. Murray, ed., *A New English Dictionary on Historical Principles . . .* [*The Oxford English Dictionary*] (Oxford, 1888), I, 454; Johnson, *Dictionary of the English Language,* I, n.p. For examples of Americans of the late 18th century using "army" in its broad sense see Rufus Putnam, "Thoughts on a Peace Establishment for the United States . . ." (1783), in Rowena Buell, ed., *Memoirs of Rufus Putnam* (Boston, 1903), 208; [Henry Knox], *A Plan for the General Arrangement of the Militia of the United States* (New York, 1786), 20, 31; Maj. Gen. William Shepard to Governor Bowdoin, Jan. 26, 1787, in "Documents Relating to the Shays Rebellion, 1787," *American Historical Review,* II (1897), 694; *Annals of Congress,* II, 1824, III, 337-340; Report of Court of Inquiry on General Harmar, Sept. 24, 1791, in *American State Papers: Military Affairs* (Washington, D. C., 1832), I, 20-35.

45. There are some passing uses of "army" in its broad sense in the literature of the ratification controversy, but these do not occur in explications of the armies clause. See, for example, "The Letters of a Landholder," in Ford, ed., *Essays,* 15; and *The Federalist,* No. 29, 181.

46. [Noah Webster], "An Examination into the Leading Principles of the Fed-

compulsory under the Constitution, just as it had been earlier, they assigned considerable importance to clarifying control over the militia. This is evidenced not only by the detail in Article I, Section 8, and in the Second Amendment, but also by the extensive debates on the problem during the ratification controversy.[47] The crucial question is whether these debates indicate that Americans of 1787-1788 believed that the militia obligation would be the sole source of compulsory military service.

The two sides characteristically disagreed about the extent of the proposed government's control over the state militias, with the Antifederalists describing federal control in more sweeping and menacing terms. By virtue of federal power to call out the militia on specified occasions and to regulate and discipline it, the new government, claimed the Antifederalists, might use the militia to destroy political liberty—that is, to subvert republican government both nationally and, what was seen as a greater threat, within the states. The charge was similar to the Antifederalist argument against a federal standing army. In several comments on the militia, however, the Constitution's opponents went revealingly beyond their attack on armies: the militia, subject to federal control, might pose a threat not only to political but to personal liberty.[48] An extended quotation from the dissent of the minority in

---

eral Constitution . . . ," in Ford, ed., *Pamphlets,* 52; Elliot, ed., *Debates,* III, 90, 391-392, 400. See also *ibid.,* II, 537. "Executing the law" may appear to be a loose category, as indicated by the argument that any use of militia authorized by a law is an execution of the law authorizing its use. Americans in the 1780s and 1790s, however, seem more specifically to have equated "to execute the Laws" with enforcing the law and suppressing opposition to law and the lawful operations of government, as in Massachusetts in 1786-1787. See *The Federalist,* No. 28, 178-179; Elliot, ed., *Debates,* III, 390; *Letters from the Federal Farmer,* 26; Farrand, ed., *Records,* I, 21, 245, III, 118; *United States Statutes at Large,* I, 262 (Act of May 2, 1792, Sec. 2), 424 (Act of Feb. 28, 1795, Sec. 2).

47. A convenient collection of statements from the debates is James Scott Brown, "The Militia," *Senate Document* No. 695, 64th Cong., 2d sess. (Washington, D. C., 1917). Some of Brown's bracketed inserts clarifying speakers' allusions to other speakers are incorrect.

48. "Letters of Centinel," in McMaster and Stone, eds., *Pennsylvania and the Federal Constitution,* 585-588; "The Address and Reasons of Dissent of the Minority of the [Pennsylvania] Convention . . . ," *ibid.,* 480-481; Martin, "Genuine Information," in Farrand, ed., *Records,* III, 207-209; "Report of the Committee on Amendments of the Maryland Convention," in Elliot, ed., *Debates,* II, 552.

the Pennsylvania convention concretely illustrates the difference in the treatment of the two military forces:

> A standing army in the hands of a government placed so independent of the people, may be made a fatal instrument to overturn the public liberties; it may be employed to enforce the collection of the most oppressive taxes, and to carry into execution the most arbitrary measures. An ambitious man who may have the army at his devotion, may step up into the throne, and seize upon absolute power.
>
> The absolute unqualified command that Congress have over the militia may be made instrumental to the destruction of all liberty, both public and private; whether of a personal, civil or religous nature.
>
> First, the personal liberty of every man, probably from sixteen to sixty years of age, may be destroyed by the power Congress have in organizing and governing of the militia. As militia they may be subjected to fines to any amount, levied in a military manner; they may be subjected to corporal punishments of the most disgraceful and humiliating kind; and to death itself, by the sentence of a court martial. To this our young men will be more immediately subjected, as a select militia, composed of them, will best answer the purposes of government.
>
> Secondly, the rights of conscience may be violated, as there is no exemption of those persons who are conscientiously scrupulous of bearing arms. . . .
>
> Thirdly, the absolute command of Congress over the militia may be destructive of public liberty; for under the guidance of an arbitrary government, they may be made the unwilling instruments of tyranny.[49]

Clearly, the Pennsylvania minority considered militia service to be compulsory. Less obviously but more important, their protest implies that army service is voluntary. While the army emerges as "a fatal instrument to overturn the public liberties," there is no indication that it would act unwillingly, as would be the case with a federalized militia. This, however, may indicate only a presumption that armies corrupted their members and not a recognition that they legally rested on voluntary recruitment. More compelling evidence flows from the Pennsylvanians' failure to charge the army

49. "The Address and Reasons of Dissent," in McMaster and Stone, eds., *Pennsylvania and the Federal Constitution,* 480-481.

with endangering personal liberty, which contrasts with their detailed charge against a federally controlled militia in this regard. Evidently, the citizen could not be forced to submit involuntarily to army discipline. Pointing significantly to the same conclusion is George Mason's comment that under the Constitution "such severities might be exercised on the militia as would make them [the freemen] wish the use of the militia to be utterly abolished, and assent to the establishment of a standing army."[50]

Still, these Antifederalists may merely have assumed that compulsory army service, although permissible under the armies clause, was not a prospect serious enough to deserve comment. They may have concluded that an army draft was unlikely, and that if it did occur, it would affect far fewer individuals than did compulsory militia duty. Yet, to reiterate a point already made, Americans had experienced the unpleasantness of a military draft, and the Antifederalists could have effectively painted a bleaker picture of the Constitution had they portrayed the armies clause as carrying even the remote possibility of compulsory service. Moreover, had they thought such a portrayal plausible, they probably would have made it, for as Cecelia Kenyon has observed, "they probed the proposed new government for every conceivable weakness or defect."[51]

It is risky, of course, to derive the original understanding of the Constitution from the views of the losers in the contest over it. What gives merit to the approach in this instance is that the view in question—that is, the basis of service in the militia and army—was not the point at issue among the contestants of 1787-1788. Rather, their controversy was over the constitutional uses to which an army or militia might be put. So, while the evidence is too murky to allow certainty, the Antifederalists' fears regarding armies and militia strongly suggest that Americans of the day believed that the citizen's militia obligation would be the sole source of compulsory service under the Constitution.

Several Federalist remarks imply a similar outlook. The view emerges, albeit not so strongly, from Hamilton's contrast between

50. Elliot, ed., *Debates*, III, 402.

51. Kenyon, ed., *The Antifederalists*, xxxviii. See also Jackson Turner Main, *The Antifederalists: Critics of the Constitution, 1781-1788* (Chapel Hill, N. C., 1961), 134.

a militia and an army with reference to the burdens each would place on the people.[52] It may have underlain the fact that although Edmund Randolph twice argued that there was no reason to fear the militia–even if it were under a certain degree of federal control–because the militia comprised the people themselves, he did not make a similar argument in defense of the army.[53] (At a later day, advocates of conscription were to argue that draftees provided a healthy civilian presence in the army.) Particularly noteworthy is a comment by Wilson Nicholas in the Virginia ratifying convention. Even if a militia were adequate for national defense, contended Nicholas, it would place unequal burdens on the people. "The poor man does as much as the rich," he said, and then asked, "Is this just? What is the consequence when war is carried on by regular troops? They are paid by taxes raised from the people, according to their property; and then the rich man pays an adequate share."[54] Nicholas's remark most readily makes sense if one assumes that while militia service is compulsory and thus falls indiscriminately on rich and poor, army service would be voluntary. The same understanding seems to have stood behind an argument, advanced by Hamilton, that derived lessons from Shays's Rebellion and from the Wyoming Valley unrest in Pennsylvania. Occasionally a rebellious spirit may so generally infect the people, Hamilton cautioned, that government cannot rely on the militia–drawn, of course, from the people–to maintain order. In such circumstances, an army, which he called a "force different from the militia," had an internal function.[55]

The fact remains, however, that within the decade preceding the Constitution's ratification, Americans had experienced com-

52. See *The Federalist,* No. 24, 156-157.

53. See Farrand, ed., *Records,* II, 387; Elliot, ed., *Debates,* III, 401. See also Alexander Contee Hanson, "Remarks on the Proposed Plan of a Federal Government . . . ," in Ford, ed., *Pamphlets,* 235-236.

54. Elliot, ed., *Debates,* III, 389.

55. *The Federalist,* No. 28, 177. The available information on the legal basis of the Massachusetts army during Shays's Rebellion is sketchy and scattered, and this is even more true of the Pennsylvania force. Some evidence may be gleaned from Massachusetts Archives, MS, Vol. 189, fols. 58-63, 68-69A, Statehouse, Boston; Jeremy Belknap to his wife, Jan. 14, 1787, in "Belknap Papers," Massachusetts Historical Society, *Collections,* 6th Ser., IV (1891), 325-326; *Resolves of the General Court . . . of Massachusetts . . . February 1787*

pulsory military service. During the War for Independence, when recruiting lagged, Washington urged Congress to ask the states to draft men for their Continental battalions. Responding in February 1778, Congress resolved "that the several states . . . be required forthwith to fill up by drafts from their militia, [or in any other way that shall be effectual,] their respective batallions." Some states had already resorted to militia drafts, and others soon complied with the congressional request.[56]

In 1787, when Randolph in Philadelphia and Hamilton in New York recalled the unhappy experience of the Revolutionary draft,[57] their recollections were not faulty. During the war, states readily sought refuge in the "in any other way" clause of the congressional resolution. To avoid outright drafting they commonly increased monetary and land bounties for volunteers, provided exemption from service for any two men who produced an able-bodied recruit, formed the militia into classes that would then hire their own recruits, offered pardons to deserters who returned to service, and allowed individuals to hire substitutes. The draft measures, as they evolved, further reflected the hostility to compulsion. After initial acts were passed, not only did subsequent acts prove necessary when men failed to materialize, but these acts

---

(Boston, 1787), 183 (Resolve of Feb. 8, 1787); Robert J. Taylor, *Western Massachusetts in the Revolution* (Providence, R. I., 1954), 158-159; "Message from the President of Pa. and Supreme Executive Council to the General Assembly," Oct. 26, 1787, *Minutes of the Supreme Executive Council of Pennsylvania . . . from July 4th, 1787, to February 6th, 1789 . . .* (*Colonial Records of Pennsylvania,* XV [Harrisburg, 1853]), 304; "Minutes of Council meeting of Oct. 30, 1787," *ibid.,* 307.

56. Fitzpatrick, ed., *Writings of George Washington,* X, 366; *Jour. Cont. Cong.,* X, 200 (Resolution of Feb. 26, 1778). Numerous examples of state legislation for raising troops may be found in Selective Service System, *Backgrounds of Selective Service,* II, Pts. 2-14, but while this purports to be a complete compilation, it omits a number of measures. I have supplemented it with a review of recruiting and draft acts and resolves which were published during the years 1776-1781 either separately or in collections of state statutes, as found in the Early American Imprint Series, microcard edition. The conclusions in the following three paragraphs are based on this legislation as supplemented by Arthur J. Alexander, "How Maryland Tried to Raise Her Continental Quotas," *Maryland Historical Magazine,* XLII (1947), 184-196; Russell F. Weigley, *History of the United States Army* (New York, 1967), 41-42; Higginbotham, *War of American Independence,* 392-393; and Hugh F. Rankin, *The North Carolina Continentals* (Chapel Hill, N. C., 1971).

57. Farrand, ed., *Records,* I, 19, 25; *The Federalist,* No. 22, 138.

imposed stiffer penalties on officials who failed to conduct drafts or to procure the necessary men.[58]

One would think, then, that memories of the draft should at least have raised questions, despite the prevalent opinion that direct federal power to recruit and tax would itself produce an adequate army. The probable explanation for the silence of the Antifederalists on this matter is that it was recognized that the wartime draft had invariably been a militia draft. When other means of manpower procurement failed, shortages were typically met by dividing the state or local militia into as many classes as there were deficiencies, with one man being chosen by lot from each class. Those selected would serve in units composed of others similarly drafted or with regular units of the state lines.[59] In one instance, in Rhode Island, a draft law drew from "all male Persons whosoever, of the Age sixteen years and upwards . . .

58. For example, *Resolves of the General Court of . . . Massachusetts . . . begun . . . on . . . the Twenty-First Day of October . . . 1780* (Boston, 1781), 40 (Resolve of Dec. 2, 1780); *ibid.*, 122 (Resolve of Feb. 26, 1781). *Resolves of the General Court of . . . Massachusetts . . . begun . . . on . . . the Thirtieth Day of May . . . 1781* (Boston, 1781), 15 (Resolve of June 16, 1781), 95 (Resolve of Oct. 20, 1781); William Waller Hening, ed., *The Statutes at Large; Being a Collection of All the Laws of Virginia . . .* (Richmond, 1819-1823), IX, 334-349 (Act of Oct. 20, 1777), 591-592 (Act of Oct. 5, 1778), X, 82-83 (Act of May 3, 1779), reprinted in Selective Service System, *Backgrounds of Selective Service,* II, Pt. 14, 344-356, 361-362, 363-364.

59. Maryland, for example, provided in its 1778 act that "the men draughted, together with the officers sent from this state, shall not be turned into companies with the regular soldiers, unless by their own choice or voluntary act, and shall be subject only to such rules and regulations as shall be agreed upon by the General Asembly of Maryland." *Laws of Maryland . . . ,* Mar. Sess., 1778 (Annapolis, 1778), Chap. V, Sec. XVI. This was not typical, however, and the state's 1780 act accorded with the more usual practice, insofar as the wartime legislation reveals practice, in ordering that a draftee "shall be from thenceforth, to every intent and purpose, considered as an inlisted soldier." *Ibid.,* Oct. Sess., 1780 (Annapolis, 1780), Chap. XLIII, Sec. VI. For similar statutes see, for example, Massachusetts House of Representatives, Resolution of Apr. 30, 1777 (Boston, 1777), Early American Imprint Series, Evans No. 15432; Hening, ed., *Statutes at Large,* X (1822), 82 (Act of May 3, 1779), in Selective Service System, *Backgrounds of Selective Service,* II, Pt. 14, 363. The distinction between regulars and draftees was usually maintained in their respective terms of service. Congress's original draft recommendation was for a nine-months' term. The terms of service enacted by the states varied considerably but generally fell short of the three-years-or-the-duration favored, but not always enforced, for enlisted regulars.

(Indians, Mulattoes and Negroes excepted)."[60] A close reading of the entire act reveals, however, the assumption that such "male Persons" were members of the militia.[61]

The pattern during the Revolutionary War is clear: compulsory service, although it existed, was based on the militia. Moreover, even when drafted militiamen served with the army, they nevertheless were fighting in a conflict which was easily associated with the repelling of invasion, for which the militia could lawfully be used. Thus recollections of the war probably did not contradict American assumptions in the late 1780s about compulsory military service but rather reinforced them.

The pattern, and the corresponding assumption about the voluntary nature of army service, also appeared in plans for a military establishment that were advanced during the Confederation period. At the request of a committee of Congress headed by Hamilton, Washington submitted his "Sentiments on a Peace Establishment" in May 1783. In discussing the nation's militia he took "as a primary position and the basis of our system, that every Citizen who enjoys the protection of a free Government, owes not only a proportion of his property, but even of his personal services" to it. In outlining a regular force he proposed that the men composing it be enlisted and said nothing to imply that they would be raised in any other way; his strong implication was rather that pay was the usual and customary way of obtaining regular troops. The committee's resulting report to Congress was consistent with Washington's "Sentiments" on these points, and the "Sentiments" themselves reflected similar, if less developed, recommendations which he had received from members of his staff.[62]

60. "An Act for filling up and compleating this State's Quota of the Continental Army," *At the General Assembly of the Governor and Company . . . Session Laws,* Nov. Sess., 1780 (Providence, R. I., 1780), 35.

61. See esp. *ibid.,* 40, 42.

62. Fitzpatrick, ed., *Writings of George Washington,* XXVI, 374-398 (esp. 383-384, 389); *Jour. Cont. Cong.,* XXV, 722-744 (esp. 736-737, 741-742). See The George Washington Papers, 4th Ser., Library of Congress (microfilm ed., Reel 91), for the staff recommendations, the most detailed of which came from Baron von Steuben, Rufus Putnam, and Timothy Pickering. The latter two letters are more conveniently available in Putnam, "Thoughts on a Peace Establishment,"

The following year, a member of Washington's wartime staff, Gen. Friedrich von Steuben, published his *Letter on the Subject of an Established Militia and Military Establishments.* In discussing "an established Continental Corps" Steuben, too, evidently simply assumed that regulars would be enlisted. Regarding the militia, however, he recognized that compulsion was its legal base in America, but he doubted the military efficacy of compelling service and hence proposed another basis for his reserve force. "The best method of engaging men in service," he wrote concerning his militia "legions," "is bounty; the expence will not be great and the inconveniency and ill humour which attends draughting [will] be avoided."[63] Steuben thus considered the problem of compulsory service important enough to make clear his difference with his former commander-in-chief over the proper basis for a militia. Yet despite his doubts about drafting, he saw no need to raise the issue in his discussion of a regular force. This bolsters the inference that he took as an unquestioned premise the voluntary basis of army service.

Alternatively, Washington and Steuben may each have ignored the option of drafting into the army because they thought it unconstitutional under the Articles of Confederation, and not simply because they took a volunteer army for granted. It is unlikely, however, that this explains their positions, for in other areas they did not diplay similar restraint. Neither man was reluctant to offer suggestions respecting the militia (an extremely touchy subject) that would have required either modification of the Articles or

---

in Buell, ed., *Memoirs of Putnam,* 198-215 (esp. 208-213), and Pickering, "Thoughts on the Military Establishment Proper for the United States at the Conclusion of the War in 1783," in Charles W. Upham, *The Life of Timothy Pickering,* IV (Boston, 1873), 431-443 (esp. 434-435, 438). See Kohn, "The Federalists and the Army," 93-132, and Russell F. Weigley, *Towards an American Army: Military Thought from Washington to Marshall* (New York, 1962), 10-16.

63. Friedrich von Steuben, *A Letter on the Subject of an Established Militia and Military Establishments . . .* (New York, 1784), esp. 2-6, 14. The other major plan offered during the Confederation period—that of Secretary at War Knox in 1786—pertained only to the militia. It recognized the principle of compulsory service in the militia, but because it did not discuss a regular army, it is not particularly revealing with respect to the distinctions treated in this article. See [Henry Knox], *A Plan for the General Arrangement of the Militia of the United States* (New York, 1786), esp. 7-10. On the Confederation period generally see Ward, *Department of War,* 1-92.

some form of independent state approval. Also, both generals called for an army free from state ties, which contradicted existing practice and would have required, at minimum, modification of the provision of Article IX that state legislatures "shall appoint the [army's] regimental officers," and probably modification of the same article's provision that the state legislatures "shall . . . raise the men."

Besides heeding debate at home, educated and politically aware Americans doubtless knew about foreign recruiting methods and specifically about the English military experience.[64] In this connection it should be recalled that Hamilton found in the Constitution the power to raise military forces "in the customary and ordinary modes practiced in other governments."[65] It is significant, then, that before the wars of the French Revolution and of Napoleon, which still lay in the future, "the customary and ordinary modes" of raising armies in Europe did not include legal compulsion, although continental nations had sporadically used it.[66]

Most important, army service in England rested in law on voluntary enlistment, the major exception being the impressment of vagrants.[67] Vagrants were also pressed in America during the

64. On the American interest in foreign developments see Michael Kraus, *The Atlantic Civilization: Eighteenth-Century Origins* (Ithaca, N. Y., 1949); Felix Gilbert, *To the Farewell Address: The Beginnings of American Foreign Policy* (New York, 1965). The evidence reviewed in these studies, with that in H. Trevor Colbourn, *The Lamp of Experience: Whig History and the Intellectual Origins of the American Revolution* (Chapel Hill, N. C., 1965), confirms John Jay's opinion in 1787 that "the history of Great Britain is the one with which we [Americans] are in general best acquainted, and it gives us many useful lessons." *The Federalist,* No. 5, 24.

65. *The Federalist,* No. 23, 149. See also James Wilson's comment: "I have taken some pains to inform myself how the other governments of the world stand with regard to this power [of raising and maintaining standing armies] . . . ." Elliot, ed., *Debates,* II, 520.

66. See, for example, Spencer Wilkinson, *The French Army before Napoleon* (Oxford, 1915), 84-86; William O. Shanahan, *Prussian Military Reforms, 1786-1813* (New York, 1945), 36-56; Theodore Ropp, *War in the Modern World,* rev. ed. (New York, 1962), 53-57; Elbridge Colby, "Conscription in Modern Form," *Infantry Journal,* XXXIV (1929), 645-646, 651; Elbridge Colby, "Conscription," in *Encyclopedia of the Social Sciences* (New York, 1935), IV, 220-222; and works cited in nn. 67, 69 below.

67. Charles M. Clode, *The Military Forces of the Crown: Their Administration and Government* (London, 1869), I, 39, II, 1-27; R. E. Scouller, *The Armies of Queen Anne* (Oxford, 1966), 106-115; Correlli Barnett, *Britain and Her Army,*

colonial period and the War of Independence, but since Englishmen and Americans defined vagrancy as a crime, this service constituted punishment, comparable in law to putting vagrants to work in other ways.[68] English governments from Elizabeth's reign through the Civil War and the Interregnum had sometimes also pressed poor but nonvagrant men into army service. Both unpopular and probably contrary to law when it had been employed, this discarded English practice seems no more likely than the drafting of vagrants to have appeared to the rights-sensitive Americans of 1787-1788 as a "customary and ordinary" method of bringing law-abiding free men into the army.[69] Still earlier, in the centuries following the Norman conquest, compulsory army service in England had resulted from feudal obligations owed personally to the king. Such duty had disappeared by the eighteenth century and would hardly have applied to America.[70]

English *militia* service did remain compulsory in theory (although seldom in fact), but William Blackstone, whose *Commentaries* of 1765-1769 quickly gained an American audience, explained that under laws dating from the reign of Charles II militiamen "are not compellable to march out of counties, unless in case of invasion or actual rebellion, nor in any case compellable to march out of the kingdom."[71] This must have struck a

---

1509-1970 (London, 1970), 140-142. Although legally voluntary for nonvagrants, army enlistment in England sometimes resulted from dubious circumstances, including intoxication and perjured testimony that the alleged recruit had taken the enlistment oath. A form of duress was present, too, in provisions that freed imprisoned debtors and criminals who enlisted in the army.

68. See Richard B. Morris, *Government and Labor in Early America* (New York, 1946), 3-6, and Dorothy Marshall, *The English Poor in the Eighteenth Century* (London, 1926), 225-245. For an example of a Revolutionary War vagrant draft act see *Laws of Maryland . . .* , Mar. Sess., 1778 (Annapolis, 1778), Chap. V, Sec. VI, providing that an able-bodied man who is judged to be vagrant "shall from and after such adjudication, be considered as a soldier enlisted."

69. C. G. Cruickshank, *Elizabeth's Army,* 2d ed. (Oxford, 1966), 8-13, 25-35; Stephen J. Stearns, "Conscription and English Society in the 1620s," *Journal of British Studies,* XI (1972), 1-23, and esp. 2-3; C. H. Firth, *Cromwell's Army,* 3d ed. (London, 1921), 21, 36-37; Godfrey Davies, "Documents Illustrating the First Civil War, 1642-1645," *Journal of Modern History,* III (1931), 66, 70-71.

70. William Blackstone, *Commentaries on the Laws of England,* I (Oxford, 1765), 397-398; Cruickshank, *Elizabeth's Army,* 2-4.

71. John R. Western, *The English Militia in the Eighteenth Century* (London,

familiar note with Americans, as evidenced by the similar provision in the Constitution and by the colonial practice of using militia for defense and volunteers for offensive expeditions.[72]

Several of the amendments suggested by the states during the ratification process are revealing. From Virginia, North Carolina, and Rhode Island came proposals that men who had religious scruples against war should be exempted from military duty upon payment for a substitute.[73] These proposals testify to the widespread understanding that compulsory military service of some sort would be permissible under the Constitution. The question again is, what sort? Although the issue of conscientious objection does not appear in the records of the ratification debates in the three states that proposed recognition of the right, the previously quoted remarks of the Pennsylvania minority suggest the fear underlying these amendments: through its power over the militia the federal government might harass men of tender conscience.

Evidence from the unsuccessful attempt to include a conscientious objector clause in the Bill of Rights provides corroboration. Madison's proposal in 1789 for what would become the Second Amendment was that "the right of the people to keep and bear arms shall not be infringed; a well armed and well regulated militia being the best security of a free country: *but no person religiously scrupulous of bearing arms shall be compelled to render military service in person.*" The context establishes that compulsory *militia* service for conscientious objectors was Madison's concern in the clause here italicized. In addition, subsequent debate in the First Congress indicated that others, whatever their views of the merits of Madison's amendment, assumed that the clause

---

1965); Blackstone, *Commentaries*, I, 399; Clode, *Military Forces of the Crown*, I, 31, 33-34, 45-46.

72. Shy, *Toward Lexington*, 3-17. Shy notes that the system of "militia for protection [and] volunteers for expeditions . . . fail[ed] at the beginning of the Seven Years War." *Ibid.*, 17. This refers, however, not to an eradication of the distinction but to the introduction of British regulars to supplant both militia and colonial volunteers. Shy also notes the renewed American praise for the militia system in the early 1770s. *Ibid.*, 377-378.

73. Elliot, ed., *Debates*, I, 335, III, 659, IV, 244.

pertained to the militia.[74] Madison and his colleagues evidently believed that such a provision was unnecessary with respect to armies; this most likely implies that they did not think service in armies could be compulsory.

An alternative explanation is that through his amendment Madison sought to prevent the *states* from harassing conscientious objectors by subjecting them to militia duty. He did not think that any guarantee was needed with respect to federal army service, not because army service could only be voluntary but because the federal government could be trusted to exempt those with religious scruples against service.[75] This argument is not persuasive. In his complete proposal Madison displayed an ability to make quite specific those restraints he wished to put on the states. He called for adding a free press guarantee both to Article I, Section 9 (as a restraint on the federal government), and to Article I, Section 10 (as a restraint on state governments). He included his militia-conscientious objector clause only in his suggested modifications to Section 9.[76] Thus he apparently feared that the federal government, unless further restrained, might force conscientious objectors into militia service. But it makes little sense to argue that Madison thought that a government with such inclinations could nevertheless be trusted not to compel army service from conscientious objectors. The more reasonable explanation is simply that he thought the federal government could not compel army service at all.

Also among the state proposals were requests from Virginia, North Carolina, and a committee of the Maryland convention for an amendment to limit enlistments to four years or, during a war, to its duration. Intended to restrict standing armies, this limitation would have made little sense had the possibility of service by compulsion as well as by enlistment been open. In reference to federal control over the militia, the Maryland committee further recommended "that the militia shall not be subjected to martial

74. *Annals of Congress,* I, 434 (italics mine), 749-751, 766-767.

75. Malbin, "Conscription, the Constitution, and the Framers," 819.

76. *Annals of Congress,* I, 434-435. See ibid., 755, for evidence of Madison's awareness of the problem of guaranteeing rights against *both* state *and* federal governments. Note that Madison's proposal at this time was to incorporate his amendments *within* the body of the Constitution.

law, except in time of war, invasion, or rebellion." "All other provisions in favor of the rights of men," the committee explained, "would be vain and nugatory if the power of subjecting all men, able to bear arms, to martial law at any moment should remain vested in Congress." While its recommendations generally implied no great trust in the proposed federal government, the committee thus readily assumed that only through the militia could federal authorities involuntarily subject men to martial law.[77]

In its tardy ratification in 1790 Rhode Island proposed "that no person shall be compelled [by the federal government] to do military duty otherwise than by voluntary enlistment, except in cases of general invasion; anything in the second paragraph of the sixth article of the Constitution, or any law made under the Constitution, to the contrary notwithstanding." There are clues that this, too, had reference to militia duty. The second paragraph of Article VI, noted in the amendment, is the so-called Supremacy Clause, making the Constitution, laws, and treaties the supreme law of the land. Under the Constitution, Congress received authority to call forth the militia to enforce the law. The likely concern of the Rhode Islanders, then, was to prevent the state militia from being used involuntarily to execute federal law. Reinforcing the inference, Rhode Island's instrument of ratification provided that until the state's amendments were accepted, "the militia of this state will not be continued in service out of this state, for a longer term than six weeks, without the consent of the legislature thereof."[78]

In short, all the amendments concerning conscientious objection and other aspects of military service seem premised on the voluntary nature of armies. Because the evidence is indirect, this conclusion must remain highly tentative. Yet it is consistent with the conclusions emerging from the other evidence already reviewed.

Perhaps the strongest argument that the founders contemplated a federal power to compel service under the armies clause comes from a series of well-known and often-quoted statements typified

77. Elliot, ed., *Debates,* II, 552, III, 660, IV, 245.
78. *Ibid.,* I, 335-336.

by a comment of Hamilton in *Federalist* 23. Holding that "the authorities essential to the care of the common defence" included raising armies and providing for their support, he wrote: "These powers ought to exist without limitation: *Because it is impossible to foresee or define the extent and variety of national exigencies, or the correspondent extent and variety of the means which may be necessary to satisfy them.* The circumstances that endanger the safety of nations are infinite; and for this reason no constitutional shackles can wisely be imposed on the power to which the care of it is committed."[79] In the same vein he later expressed the opinion that "the idea of restraining the Legislative authority, in the means of providing for the national defence, is one of those refinements, which owe their origin to a zeal for liberty more ardent than enlightened."[80] Statements like these–which can be labeled "indefinite contingency arguments"–create a strong presumption that the anticompulsory service evidence needs heavy discounting.

The difficulty is that the indefinite contingency arguments must be read within both their immediate context and the general context of the debate over the military provisions of the Constitution, including other Federalist remarks and Antifederalist objections. The statement quoted above from *Federalist* 23, for example, appeared not in a discussion of drafting but in one about the merits of direct federal power as opposed to continued reliance on the states.[81] It was this problem that prompted Hamilton's opposition in the same paper to restrictions "in any matter essential to the *formation, direction,* or *support* of the NATIONAL FORCES." The immediate context establishes that he especially feared the

79. *The Federalist,* No. 23, 147. For statements see *ibid.,* No. 23, 147-148, No. 31, 195-196, No. 34, 210-211, No. 41, 270-273; and Elliot, ed., *Debates,* II, 351.

80. *The Federalist,* No. 26, 164.

81. See *ibid.,* No. 23. See also *ibid.,* No. 22, 137-138. This issue also provided the context for the other indefinite contingency statements cited in n. 79 above. A close analysis of Hamilton's remarks *ibid.,* No. 23, indicates that even if, as seems unlikely, Hamilton thought of raising armies in terms broader than those I have attributed to him, he was still not committing himself to the proposition that the federal government, as it would actually be established under the Constitution, necessarily had broader powers. He only maintained that an adequate central government "ought" to possess them. This reading seems consistent with his gloss on "ought" in the next paper. See *ibid.,* No. 24, 153n.

shackles that the quota system placed both on money and on the men whom money would attract.[82] This accords with his view, previously noted, that revenue was "the essential engine" of national defense. In short, the wartime and Confederation experiences largely defined the problem to which Hamilton was responding, and these same experiences almost certainly informed his readers' understanding of the passage.

Nevertheless, in offering their indefinite contingency arguments, Hamilton and his fellow Federalists probably had other limitations in mind as well, and these may be readily surmised. Antifederalist writers and several state conventions proposed requiring a two-thirds or three-fourths vote of each house of Congress before a standing army could be established in time of peace.[83] In the Federal Convention Luther Martin and Elbridge Gerry unsuccessfully moved that a numerical limit be placed on the army, and Gerry questioned the two-year limit on army appropriations: a one-year rule, he said, would carry greater security.[84] These two restrictions were mooted in the debates on ratification, and they, together with the question of a flat prohibition on peacetime standing armies, prompted Hamilton's comment about the dangers of excessive zeal for liberty.[85] New York proposed requiring a two-

82. *Ibid.*, No. 23, 147-149. Note, too, that Hamilton was here referring to the raising of "national forces" in "the customary and ordinary modes practiced in other governments," the implications of which are discussed above. Moreover, his reference to "national forces" rather than "armies" does not indicate that he meant something other than enlistment and professionals as regards nonmilitia land forces. Rather, his choice of words was most likely dictated by the fact that he was referring not only to armies but to the navy as well. He may also have had federalized militia in mind, but he surely did not mean to propose transgressing the Constitution's restrictions on militia calls.

83. Elliot, ed., *Debates,* I, 326, 330, II, 545, III, 660, IV, 245. See also [James Winthrop], "The Letters of Agrippa," in Ford, ed., *Essays,* 118-119.

84. Farrand, ed., *Records,* II, 330, 509. See also *ibid.,* 329. Roger Sherman of Connecticut indicated his support as well for a "reasonable" numerical limitation. *Ibid.,* 509. The proposal originally sent to the Committee of Detail did call for a one-year limitation, but the committee reported out the two-year restriction. *Ibid.,* 341, 508.

85. See, for example, *The Federalist,* Nos. 24-26; No. 41, 271-274; Elliot, ed., *Debates,* IV, 96-99; Pelatiah Webster, "Remarks on the Address of Sixteen Members of the [Pennsylvania] Assembly . . . ," in McMaster and Stone, eds., *Pennsylvania and the Federal Constitution,* 101; "Letters of Centinel," *ibid.,* 585; and Martin, "Genuine Information," in Farrand, ed., *Records,* III, 207.

thirds vote of Congress for declarations of war and wished to prohibit the president from personally commanding troops in the field without the consent of Congress. North Carolina would have prevented Congress from bringing foreign troops into the country without a two-thirds vote of each house.[86]

Thus when the Federalists warned against limiting the means available to the federal government for raising armies, they and their contemporaries probably understood quite well what restrictions they wished to avoid: limitations, in favor of the states, on direct federal authority to raise troops and revenue; a prohibition on standing armies during peace; a requirement for qualified majorities as a prerequisite to maintaining peacetime armies; and several less frequently mentioned restrictions. The Federalists were not attacking a prohibition on federal authority to draft under the armies clause; neither were the Antifederalists proposing it. Both sides knew what the Constitution meant by "armies" and so had no reason to discuss an army draft. Thus, too, both sides would probably have rejected as self-contradictory the argument advanced in later years that the necessary-and-proper clause of Article I, Section 8, made drafting a permissible means of implementing the armies clause.

In establishing a new government Americans held conflicting aspirations regarding the military. They valued both defense and liberty; they sought both military security and guarantees against the tyranny they associated with military establishments. These conflicts evoked substantial debate at Philadelphia and in the states, yet the contestants said little about the issue of compulsory military service, which would take on such significance for millions in later generations. Still, sketchy as it is, the evidence allows fairly specific conclusions.

Americans in 1787-1788 accepted compulsory military service: they had lived with it for years. They nevertheless took a particular view of it, which they probably read into the Constitution. In their experience, the militia had been the sole basis for compulsory service, and this had been true even during the recent War

86. Elliot, ed., *Debates,* I, 330, IV, 247. Note, too, the proposed limitations on terms of enlistment, cited in n. 77.

of Independence. "Armies," as the term was used in the Constitution, comprised professional troops serving by legally voluntary enlistment and motivated by financial reward. The Constitution greatly improved on the Articles of Confederation by authorizing the central government directly to enlist and maintain regular troops and to raise funds for attracting and supporting them, without reliance on the states. These powers, and not the power to draft regulars, were thought crucial in providing federal authorities with the means necessary to meet unforeseen contingencies. In addition, while the state militias would remain a bulwark of liberty (and the Second Amendment was soon to reinforce this guarantee), the federal government could prescribe standards for them and had carefully controlled access to them. Accordingly, through their militia obligation men might find themselves serving involuntarily in what could be called a federal "army," applying the term in the broad, nontechnical, and nonconstitutional sense in which it was then sometimes used, but such service would be limited to executing the laws, suppressing insurrections, and repelling invasions.[87]

87. These conclusions have not controlled twentieth-century constitutional interpretation by the courts. To forestall misunderstanding of the purpose of this article—and to oversimplify a complex and very different issue from the one herein discussed—I should offer my view that changed conditions and new concepts of citizenship introduced by the Fourteenth Amendment provide considerable, although perhaps not complete, constitutional warrant for the result reached in the *Selective Draft Law Cases* of 1918.

3

THE ORIGINS OF THE TENTH AMENDMENT: *History, Sovereignty, and the Problem of Constitutional Intention*

I. Fire Amidst Constitutional Ashes: The Search for Historical Meaning

Why examine the Tenth Amendment? An authoritative survey of Supreme Court decisions through 1972 found no practical force to the Amendment's guarantee that "[t]he powers not delegated to the United States by the Constitution, nor prohibited by it to the States, are reserved to the States respectively, or to the people."[1] The Amendment tellingly received only eight pages in the massive volume, against 140 for the commerce clause and 255 for the Fourteenth Amendment, each of which has had obvious constitutional importance in limiting state discretion.[2]

More recently, however, stirrings have been visible. In 1975, *Fry v. United States* announced by way of dictum that the Tenth

1. See Congressional Reference Service, Library of Congress, The Constitution of the United States of America: Analysis and Interpretation 1263-71 (1973).

2. See *id., passim.*

Amendment "is not without significance. . . . [It] expressly declares the constitutional policy that Congress may not exercise power in a fashion that impairs the States' integrity or their ability to function effectively in a federal system."[3] The next year, speaking for the Court in *National League of Cities v. Usery,* Justice William H. Rehnquist relied on the Amendment, as well as on limits which arise from "our federal system of government" and to which Amendment gives "an express declaration," to overturn federal wages and hours legislation as applied to state and municipal employees.[4] Some sense of the controversial quality of the close decision in *Usery* emerges from Justice William J. Brennan's dissent, which found in Rehnquist's opinion a "portent . . . so ominous for our constitutional jurisprudence as to leave one incredulous."[5]

Coming after forty years of desuetude for the Tenth Amendment, *Usery* reaffirms the Amendment's constitutional potential. In broader historical terms, of course, this potential is readily apparent. From the Taney Court through the heyday of the "Four Horsemen" in the 1930s, the Amendment was associated with "Dual Federalism." This restrictive doctrine rested on the propositions that the Constitution had not vested the federal government with sovereign powers of internal police, and that by the overall constitutional scheme–a scheme "made absolutely certain by the Tenth Amendment"–those sovereign powers were retained by the states and thus positively prohibited to the central government.[6] By 1950, to be sure, Edward S. Corwin proclaimed "the passing of dual federalism" and wrote, with considerable accuracy, that "what was once vaunted as a Constitution of Rights, both State and private, has been replaced by a Constitution of Powers."[7]

3. Fry v. United States, 421 U.S. 542, 547 n. 7 (1975) [hereinafter *Fry*].

4. National League of Cities v. Usery, 426 U.S. 833, esp. 842-44 (1976) [hereinafter *Usery*].

5. *Id.* at 875. See generally, *e.g.*, Barber, *National League of Cities v. Usery: New Meaning for the Tenth Amendment?* 1976 SUP. CT. REVIEW 161; Lofgren, *National League of Cities v. Usery: Dual Federalism Reborn,* 4 CLAREMONT J. OF PUBLIC AFFAIRS 19 (Spring 1977).

6. See *e.g.*, E. CORWIN, THE COMMERCE POWER VERSUS STATES RIGHTS, *passim* (1936), which excerpts the leading cases. The quotation in the text is from Kansas v. Colorado, 206 U.S. 46, 89 (1907).

7. Corwin, *The Passing of Dual Federalism,* 36 VA. L. REV. 1, 2 (1950).

During its dormancy in the four decades following *National Labor Relations Board v. Jones and Laughlin Steel Corporation,*[8] the Amendment nevertheless found an unlikely champion. In fact, Justice William O. Douglas's occasional arguments for state sovereignty, which he claimed was "attested by the Tenth Amendment," anticipated those advanced by the Court in *Usery.*[9] Finally, even Justice Brennan's dissent in the case allowed that the Amendment supported some role for the individual states within the constitutional system.[10]

For Justice Brennan, though, the Amendment's function was far more symbolic than substantive; he appropriated–and challenged the majority with–the historical gloss put on the Amendment in *United States v. Darby:*

> The amendment states but a truism that all is retained which has not been surrendered. There is nothing in the history of its adoption to suggest that it was more than declaratory of the relationship between the national and state governments as it had been established by the Constitution before the amendment or that its purpose was other than to allay fears that the new national government might seek to exercise powers not granted, and that the states might not be able to exercise fully their reserved powers.[11]

Justice Rehnquist, who contrarily saw the Amendment as having independent force, neither quoted it nor explored its history in *Usery,* but he could not have quarreled with having recourse to its original understanding. Dissenting in *Fry v. United States,* he had written: "Surely there can be no more important fundamental constitutional question than that of the intention of the Framers of the Constitution as to how authority should be allocated between the National and State Governments."[12]

In short, an issue with deep historical roots has been joined in

8. 301 U.S. 1 (1937).

9. See New York v. United States, 326 U.S. 572, 590 (1946); Case v. Bowles, 327 U.S. 92, 103 (1946); Maryland v. Wirtz, 392 U.S. 183, 201 (1968) (the quotation in the text is at 205) (all dissenting opinions).

10. See *Usery* at 861 n. 4.

11. United States v. Darby, 12 U.S. 100, 124 (1941), quoted by Brennan in *Usery* at 862 (Brennan's italics omitted).

12. *Fry* at 559.

a modern setting. It is accordingly worthwhile to attempt to clarify the ambiguities surrounding the Amendment so far as they are susceptible to clarification through a reconstruction of its origins and "original understanding." Such an effort may also clarify why ambiguities almost necessarily remain—that is, why resolving certain historical issues only opens other questions. And it seems appropriate in this context to make a few preliminary and somewhat theoretical remarks about the kind of historical-constitutional enterprise I am undertaking.

Let me be both as unabashed and as clear as I can: the historical issues herein pursued do take on *added* interest because they have figured in a recent constitutional dispute of some significance. But a caveat is in order; this added interest remains in a sense a *secondary* interest for the historian. The issues are primarily interesting because they are there. They offer the historian puzzles about people who acted as if, and evidently thought, that they were up to important things, and who, for reasons which could be adduced, in retrospect seem really to have been engaged in critical activities. This line of comment, I realize, could be unfolded and developed at length into an area different in emphasis from the topic of the papers in this collection, for it easily leads into fundamental questions about the nature, purposes, and limitations of the historical enterprise. I have no intention here of pursuing these questions,[13] except to mention a difference, at least in degree, between the historian's goal and what I take to be the lawyer's. As a historian, I feel no obligation to produce *an* answer to the question of what the Tenth Amendment originally meant. A comment by C. H. McIlwain is apposite: "No single fault has been the source of so much bad history as the reading back of later and sharper distinctions into earlier periods where they have no place."[14]

I imagine there are few today anyway who would look exclusively to the 1780s to determine the present significance and

13. The literature on the subject is of course enormous; see, *e.g.*, E. Carr, What is History? (1961); D. Fischer, Historians' Fallacies: Toward a Logic of Historical Thought (1970); A. Marwick, The Nature of History (1970).

14. C. McIlwain, The American Revolution: A Constitutional Interpretation 64 (1923).

meaning of the Constitution or any of its early amendments. As Justice Holmes reminded us, "when we are dealing with words that are also a constituent act, . . . we must realize that they have called into life a being the development of which could not have been foreseen completely by the most gifted of its begetters." "Our whole experience," he cautioned, needs consideration and "not merely" what was said during the founding period.[15] Still, Holmes's choice of the phrase "not merely" is worth pondering. It suggests that one should at least take account of accurate history in fashioning constitutional arguments, even if that history itself is not fully dispositive of the issues in question. So the lawyer and the historian share a common interest. Certainly Justice Robert Jackson's remark in another context is not *always* true: we are not invariably forced to divine what our forefathers envisioned "from materials almost as enigmatic as the dreams Joseph was called upon to interpret for Pharaoh."[16]

What sort of meaning, then, do I seek? The term "constitutional intention," used in my title, should not obscure the fact that a constitution itself can intend nothing. Rather, people give it meaning in the course of drafting, ratifying, and later interpreting it. Beyond that, because what is at issue herein is the original meaning of an amendment which became part of America's prescriptive framework of government, the understandings of greatest interest are those of the Americans who gave the Tenth Amendment its legal status.[17] Unfortunately, direct documentation for the state

15. Missouri v. Holland, 252 U.S. 416, 433 (1920).

16. Youngstown Sheet and Tube Co. v. Sawyer, 343 U.S. 579, 634 (1952) (concurring).

17. Cf. James Madison's remark in 1796 about the Constitution itself:

> "As the instrument came from them [the framers in Philadelphia] it was nothing more than the draft of a plan, nothing but a dead letter, until life and validity were breathed into it by the voice of the people, speaking through the several State Conventions. If we were to look, therefore, for the meaning of the instrument beyond the face of the instrument, we must look for it, not in the General Convention, which proposed, but in the State Conventions, which accepted and ratified the Constitution."

Quoted in 3 THE RECORDS OF THE FEDERAL CONVICTION OF 1787 at 374 (M. Farrand ed., rev. ed., 1937) [hereinafter cited as FARRAND, RECORDS]. See THE FEDERALIST, No. 40 at 263-64 (Cooke ed.) [all further citations to *The Federalist* are to the Cooke edition]; C. MILLER, THE SUPREME COURT AND THE USES OF

legislative debates over its ratification in 1789-1791 is almost non-existent.[18] At the same time, however, warranted conclusions are possible about how Americans in 1789-1791 likely understood the Amendment. After all, the Federal Convention of 1787, the state ratification debates and conventions of 1787-1788, and the discussion in Congress of James Madison's proposed bill of rights in the summer of 1789 culminated over a decade of political controversy. An exploration of pertinent aspects of this period reveals the major concerns which underlay the Amendment Congress sent to the states in September 1789 and which undoubtedly guided deliberations in the states.

II. Early Confrontations with Sovereignty and Federalism

The problem of dividing authority between two levels of government was hardly new to the late 1780s. The earlier struggle with Great Britain gave rise to the issue in pressing form and provides a useful starting point for developing the ideas which eventually shaped the Tenth Amendment.

In 1773, John Adams called the dispute with England over the proper locus of sovereignty "the greatest Question ever yet agitated."[19] William Blackstone, whose American readership was substantial, had already advanced what was the orthodox view of sovereignty in both England and America on the eve of the War for Independence. In every state, he argued, "there is and must be . . . a supreme, irresistible, absolute, uncontrolled authority, in which the *jura summi imperii,* or rights of sovereignty, reside."

HISTORY 159 n. 23 (1969). On the problems of determining constitutional intent generally, see *id., passim;* Anderson, *The Intention of the Framers: A Note on Constitutional Interpretation,* 49 AM. POL. SCI. REV. 340 (1955).

18. See THE BILL OF RIGHTS: A DOCUMENTARY HISTORY 1171 (B. Schwartz ed. 1971 (editor's note).

19. Quoted in G. WOOD, THE CREATION OF THE AMERICAN REPUBLIC, 1776-1787 at 345 (1969) [hereinafter cited as WOOD, CREATION]. Wood describes the doctrine of sovereignty as "the single most important abstraction of politics in the entire Revolutionary era," *id.,* and as "the ultimate abstract principle to which nearly all arguments were sooner or later reduced." *Id.* at 354.

In England these rights rested with the Crown in Parliament. "Sovereignty and legislature," he said, "are indeed controvertible terms."[20] Despite the assertion in the Declaration of Independence that governments "deriv[e] their just powers from the consent of the governed" and similar statements in early state declarations of rights,[21] Americans of the mid-1770s generally followed Blackstone in assigning sovereignty to the tangible organs of government, but not without displaying some confusion. Governmental power justly derived *from* the people, but the people had no means of exercising it without in the process dissolving existing governments (which Americans of course did in 1776).[22] The Massachusetts General Court stated the principle–and evidenced the confusion–in this way: "It is a maxim that in every *government* there must exist somewhere a supreme sovereign, absolute and uncontrolled power; but this power resides always in the body of the people, and it can never be delegated to one man or to a few . . . ."[23]

This conception of sovereignty influenced the break with England; it denied that Parliament and local legislatures could coexist in any but the hierarchical relationship which increasingly became anathema to Americans.[24] Because it provided no basis for the existence of an effective central government *in* America, so long as state authority persisted, the orthodox view also influenced the Articles of Confederation.

Once Independence was declared, Congress turned its attention

20. 1 W. BLACKSTONE, COMMENTARIES ON THE LAWS OF ENGLAND *46, *49-51, *160-62.

21. *E.g.*, THE FEDERAL AND STATE CONSTITUTIONS, COLONIAL CHARTERS, AND OTHER ORGANIC LAWS OF THE UNITED STATES 958 (Massachusetts 1780, Art. V), 1541 (Pennsylvania 1776, Art. IV), 1908 (Virginia 1776, Sec. 2) (B. Poore, comp. 1878) [hereinafter cited as FEDERAL AND STATE CONSTITUTIONS].

22. See WOOD, CREATION 344-63; R. ADAMS, POLITICAL IDEAS OF THE AMERICAN REVOLUTION 153-81 (1922). Blackstone himself accepted John Locke's claim that the people possessed *"a Supream Power* to remove or *alter the Legislative,"* but he also followed Locke in holding that "however just this conclusion may be in theory, we cannot practically adopt it, nor take any *legal* steps for carrying it into execution . . . ." J. LOCKE, TWO TREATISES OF GOVERNMENT, Book II, secs. 149-50, 212, 227 (P. Laslett ed. 1960); BLACKSTONE, COMMENTARIES *52, *161-62.

23. Quoted in ADAMS, *supra* note 22, at 173.

24. See, e.g., B. BAILYN, THE IDEOLOGICAL ORIGINS OF THE AMERICAN REVOLUTION 198-229 (1967).

in July 1776 to framing a plan of union.[25] On July 12, John Dickinson offered a draft which provided that "Each Colony . . . reserves to itself the sole and exclusive Regulation and Government of its internal police, in all matters that shall not interfere with the Articles of this Confederation."[26] This was the only guarantee in favor of the states in the long and detailed document, and it was a highly qualified one at that. Positive grants to Congress were admittedly limited, and most evident in the areas of war and foreign relations, but the complexities of the plan could have led to expansion through interpretation.[27] The sole explicit restriction on Congress's internal authority–a prohibition on levying taxes except for support of the post office–was eliminated in the second draft in August.[28]

Attacks on the Dickinson draft initially focused on three characteristics of the plan: equal representation of the states, apportionment of expenses on the basis of total population, and congressional control over western lands. As debate progressed, equal representation was retained, the basis for apportionment of expenses was changed to total land value, and congressional power over the west was substantially reduced.[29] But the plan's fundamental feature–the restriction of state authority to those "matters that shall not interfere with the Articles of this Confederation"–did not come in for attack until Thomas Burke of North Carolina arrived in Congress in early 1777.[30]

Just after Burke's arrival, James Wilson made several attempts on other issues to establish the principle that Congress already possessed a general authority over the states; these efforts may have sensitized Burke to the question of sovereignty.[31] Wishing in

25. For the standard account, see M. JENSEN, THE ARTICLES OF CONFEDERATION: AN INTERPRETATION OF THE SOCIAL-CONSTITUTIONAL HISTORY OF THE AMERICAN REVOLUTION, 1774-1781 esp. chs. 4-13 (1940) [hereinafter cited as JENSEN, ARTICLES].

26. 5 JOURNALS OF THE CONTINENTAL CONGRESS 546-54 (1906) (the quoted provision appears at 547).

27. See JENSEN, ARTICLES 129-37, 241-42.

28. Compare 5 JOURNALS OF THE CONTINENTAL CONGRESS 552 with *id.* at 685.

29. JENSEN, ARTICLES 140-60.

30. *Id.* at 169-70.

31. *Id.* at 170-74. The apparently catalytic issues included the permissibility of regional meetings of states without congressional approval, control over desertion

any event "that the Power of Congress [be] accurately defined" in view of "the Delusive Intoxication which Power Naturally imposes on the human Mind," Burke offered an amendment to Dickinson's plan: "all sovereign power [would be] in the States separately, and . . . particular acts of it [Congress], which should be expressly enumerated, would be exercised in conjunction, and not otherwise; but . . . in all things else each State would exercise all the rights and powers of sovereignty uncontrolled." Wilson opposed the change, but Burke's amendment prevailed by an eleven-to-one vote.[32] The resulting Article II of the Articles of Confederation confirmed the rejection of a national government of broad and potentially elastic powers; it read: "Each state retains its sovereignty, freedom and independence, and every power, jurisdiction and right which is not by this confederation expressly delegated to the united states in Congress assembled."[33]

Although a clear victory for the proponents of state sovereignty, passage and ratification of the Articles failed to end confusion over the locus of sovereignty in America.[34] Continuing disagree-

---

from the army, the basis of voting on congressional adjournments, and the nature of citizenship.

32. Letter from Thomas Burke to the Governor of North Carolina, March 11, 1777, in 2 LETTERS OF MEMBERS OF THE CONTINENTAL CONGRESS 294 (E. Burnett ed. 1923); letter from same to same, April 29, 1777, in *id.* at 345-46.

33. Art. of Confed., Art. II., in 9 JOURNALS OF THE CONTINENTAL CONGRESS 908 (1907).

34. See F. MCDONALD, THE FORMATION OF THE AMERICAN REPUBLIC, 1776-1790 at 190-91 & n. † (1967), for a summary of the positions [hereinafter cited as MCDONALD, FORMATION]. The debate over the locus of sovereignty in America in the 1770s and 1780s has continued into the twentieth century; compare Van Tyne, *Sovereignty in the American Revolution: An Historical Study,* 12 AM. HIST. REV. 529 (1907), and D. BOORSTIN, THE AMERICANS: THE NATIONAL EXPERIENCE 400-05 (1965) (both arguing state sovereignty and the priority of the states over the nation), with Nettels, *The Origins of the Union and of the States,* 58 PROC. OF MASS. HIST. SOC. 68 (1957-60), and Morris, *"We the People of the United States": The Bicentennial of a People's Revolution,* 82 AMER. HIST. REV. 1, 10-14 (1977) (both taking the opposing positions). Justice George Sutherland had it both ways, depending on whether he was discussing "internal" or "external" sovereignty. See G. SUTHERLAND, CONSTITUTIONAL POWER AND WORLD AFFAIRS 24-47 (1919); Carter v. Carter Coal Co., 298 U.S. 238 (1936); United States v. Curtiss-Wright Export Corp., 299 U.S. 307 (1936). Although the question of the devolution of sovereignty in America in the 1770s and 1780s prior to the

ment on the issue, combined with obvious limitations placed on Congress by the Articles, produced further controversies.

One of these involved the Bank of North America. James Wilson, who remained the leading exponent of congressional sovereignty,[35] defended the congressionally chartered bank when it came under attack in Pennsylvania in 1783-1785.[36] Undeterred by Article II of the Confederation, the future Associate Justice held that since the states severally had never held the power to charter a bank commensurate with the needs of the whole of the United States, they could never have retained such a power. "[T]he United States," he held, "have general rights, general powers, and general obligations, not derived from particular states, nor from all the particular states, taken separately, but resulting from the Union of the Whole . . . ." Wilson also pointed to the section of Article V "that for the more convenient management of the *general interests* of the United States, delegates shall be annually appointed to meet in Congress." Moreover, the Declaration of Inde-

---

adoption of the Constitution has thus had judicial significance, "legal" resolution of the issue is not, I think, vital to understanding the meaning of the Tenth Amendment. The issue is instead important because of the debates over it in the 1770s and 1780s and the insights into the understandings of that period which those debates provide. But see Professor Jaffa's comment: "[T]he meaning of the Tenth Amendment turns decisively upon the question of whether the states are conceived to have this prior and independent existence." Jaffa, *"Partly Federal, Partly National": On the Political Theory of the Civil War,* in A NATION OF STATES 109, 114-15 (R. Goldwin ed., 2d ed., 1974). However, if the Amendment is declaratory of the overall constitutional scheme (and Professor Jaffa apparently agrees with me that it is; see *id.* at 117-18), then the crucial question would seem to be the status of the states *under* the Constitution. But whether or not resolution of their status prior to the Constitution entails a particular answer to the question of their status under the Constitution is, in turn, a question requiring resolution of other issues (and those other issues are, roughly speaking, those debated by the people discussed in this paper). The Constitution may have formed "a more perfect Union," to take a point Professor Jaffa raises (see *id.* at 120), but even if great importance is assigned these several words, I question whether the framers and ratifiers would necessarily have accepted the proposition that the Union of the Constitution was *only* a "more perfect" (that is, in modern usage, a "more complete") version of the same *kind* of union as that in existence under the Confederation. See *infra, passim.*

35. See MCDONALD, FORMATION 191 n. †.

36. See C. SMITH, JAMES WILSON: FOUNDING FATHER, 1742-1798 at 140-68.

pendence had brought into being one nation, with all the powers of an independent nation.[37]

No single proponent of state sovereignty matched Wilson, but the bank episode shows state autonomy was a fact which the nationalists of the period had to reckon with. When Robert Morris, the Superintendent of Finance under the Confederation and no friend of state sovereignty, had proposed the Bank of North America in 1781, he questioned "whether it may not be necessary and proper that Congress should make immediate application to the several States to invest them with the power of incorporating a bank . . . ."[38] Because some delegates evidently believed Congress held the power of incorporation and others thought previous pledges bound Congress to grant the charter, the bank resolution passed, but it revealingly requested the individual states to approve the bank.[39] Noting this request, James Madison hardly anticipated his later support for a national government when he wrote: "As this is a tacit admission of a defect of power I hope it will be an antidote against the poisonous tendency of precedents of usurpation."[40]

Several attempts to amend the Articles to give Congress author-

37. Wilson, *Considerations on the Bank of North America,* 1785, in THE WORKS OF JAMES WILSON 824, 828-31 (R. McCloskey ed. 1967) (Wilson's emphasis). I believe Professor Smith is incorrect in finding in Wilson's bank defense "the first expression of the doctrine of dual sovereignty which Wilson was to develop most eloquently two years later in the Federal Convention" (see Smith, *supra* note 36, at 152), because Wilson did not develop a doctrine of *dual* sovereignty in the Federal Convention (or in the ratification debates). See *infra,* text accompanying nn. 49, 63, 68, 105-11.

38. Letter from Robert Morris to the President of Congress, May 27, 1781, in 4 THE REVOLUTIONARY DIPLOMATIC CORRESPONDENCE OF THE UNITED STATES 421 (F. Wharton ed. 1889).

39. Letter from the Virginia Delegates to the Governor of Virginia, January 8, 1782, in 6 LETTERS OF MEMBERS OF THE CONTINENTAL CONGRESS 288 (E. Burnett ed. 1933); Resolve of December 31, 1781, in 21 JOURNALS OF THE CONTINENTAL CONGRESS 1187-90 (1912). See Resolve of May 26, 1781, in 20 *id.* at 546-48 (1912); letter from Morris to the Governors of the States, January 8, 1782, in 5 REVOLUTIONARY DIPLOMATIC CORRESPONDENCE, *supra* note 38, at 94.

40. Letter from Madison to Edmund Pendleton, January 8, 1782, in 6 LETTERS OF MEMBERS OF THE CONTINENTAL CONGRESS, *supra* note 39, at 289, 290. Madison's future collaborator, Alexander Hamilton, also doubted congressional authority in this regard. See Letter from Hamilton to Robert Morris, April 30, 1781, in 2 THE PAPERS OF ALEXANDER HAMILTON 604, 629-30 (H. Syrett and J. Cooke eds. 1961).

ity over commerce and an independent power of taxation similarly evidenced recognition of the Confederation's limits. Relatedly, opposition to the proposals involved both an ever-lurking fear of consolidated government and an orthodox understanding of sovereignty.[41] As a Georgian described the reaction to a suggested commerce amendment, "The high sounding terms of sovereignty and independence have caught and inflamed the minds of some, and the dread of losing our liberty, by giving power to Congress, has intoxicated and alarmed those of others."[42] An impost amendment would give Congress the financial base which, said Abraham Yates, Jr., "is the first, nay, I say the only object of tyrants. . . . This power is the center of gravity, for it will eventually draw into its vortex all other powers."[43] It would lead to the centralized government which the wisdom of the period taught was impossible over an extended territory without loss of liberty, and which, as the Virginia legislature resolved in 1782, would be "injurious to its [the state's] sovereignty, may prove destructive of the rights and liberty of the people, and . . . is contravening the spirit of the confederation."[44]

These disputes preceding and during the Confederation period lead to several interrelated conclusions bearing on the future Tenth Amendment. First, at a theoretical level, the Blackstonian conception of sovereignty demonstrated a "staying power" making likely its continued potency in the face of a new understanding that obviated the need for explicit reservations on behalf of the states and people. Second, at a constitutional level, Americans constantly confronted Article II of the Articles of Confederation with its guarantee to the states of all powers not *expressly* delegated. Third, at a political level, controversy over the Articles themselves, the Bank of North America, a congressional commerce power, and the impost must have heightened actual awareness of the meaning, for local autonomy, of an express guarantee

41. See McDonald, Formation 133-42; M. Jensen, The New Nation: A History of the United States During the Confederation, 1781-1789 at 399-421 (1962); J. Main, The Antifederalists: Critics of the Constitution, 1781-1788 at 72-102, 110-15 (1964) [hereinafter cited as Main, Antifederalists].

42. Quoted in Main, Antifederalists 110.

43. Quoted in *id.* at 79.

44. Quoted in *id.* at 80.

in favor of the states. Most immediately, these conclusions provide a backdrop for the Philadelphia Convention.

III. The Federal Convention of 1787

The importance of the Philadelphia Convention to an understanding of the origins of the Tenth Amendment lies primarily in two areas. First, and most obviously, the Convention produced the Constitution which was debated in the states. This document had elements which predictably evoked opposition and counterproposals from Americans committed to the principles of the Articles of Confederation. Second, the discussions in Philadelphia, though kept secret at the time,[45] aid us today in further understanding then-current views about sovereignty which soon figured in the state ratification debates.

A. *James Madison's "Middle Ground" and Its Fate*

By the time the Convention met, James Madison had sketched requirements for an adequate national government. Writing George Washington, he claimed to "have sought for a middle ground, which may at once support a due supremacy of the national authority, and not exclude the local authorities wherever they can be subordinately useful." But his "middle ground" was in fact a scheme for national supremacy. Besides holding authority in all matters requiring national uniformity, his proposed central government could veto state laws *"in all cases whatsover"* and could militarily coerce the states. Also, "[t]o give [his] new System its proper validity and energy," he called for "a ratification . . . from the people, and not merely from the ordinary authority of the Legislatures."[46] As he explained to Jefferson, "such a ratification by the people themselves of the several States . . . will ren-

45. 1 Farrand, Records xi-xii. During the ratification controversy, however, some of the framers discussed their work in Philadelphia; see esp. Luther Martin's "Genuine Information . . . ," November 29, 1787, in 3 *id.* at 172-232.

46. Letter from Madison to Washington, April 16, 1787, in 2 The Writings of James Madison 344, 344-49 (G. Hunt, ed. 1901). Accord, letter from Madison to Edmund Randolph, April 8, 1787, in *id.* at 336-40. Also revealing are notes Madison made on the defects of the Confederation, and on ancient and modern confederacies, in *id.* at 361-412.

der it [the new government] clearly paramount to their Legislative authorities."[47]

When the convention met in late May 1787, Governor Edmund Randolph of Virginia introduced a plan of government embodying Madison's views.[48] Defending it, Madison, James Wilson, and others reiterated their commitments to national supremacy.[49] On one occasion Madison himself evidently went so far as to use a highly impolitic comparison: "The states ought to be placed under the control of the general government–at least as much so as they formerly were under the king and British parliament."[50] It may be a measure of the nationalists' zeal that in their despondency over defeat of proportional representation in the Senate they apparently overlooked what they had gained in the Supremacy Clause.[51]

Proposing a broad national plan was one thing; securing its adoption was another. As John Roche has aptly observed of the framers, "For over three months, in what must have seemed to the faithful participants an endless process of give-and-take, they reasoned, cajoled, threatened, and bargained amongst themselves."[52] In the end, Congress had no general power of legislation, no veto over state laws, and no power of coercion.

Still, once state equality in the Senate was secure, little opposition remained to a strong national government.[53] After defeating

47. Letter from Madison to Thomas Jefferson, March 18, 1787, in *id.* at 324, 326.

48. 1 FARRAND, RECORDS 18-23 (Madison's notes, May 29).

49. See, *e.g., id.* at 36-37, 136, 282-93, 323, 492; 2 *id.* at 390-92 (all Madison's notes, May 30, June 6, 18, 19, 30, August 23).

50. 1 *id.* at 471 (Yates's notes, 29 June). In his old age, Madison tried to explain away this comment and his accompanying remarks about sovereignty quoted *infra,* text accompanying n. 61, but his own notes indicate he probably made the comments. See letters from Madison to N.P. Trist, December [?], 1831, and to W.C. Rives, October 21, 1833 in 3 *id.* at 516, 521; 1 *id.* at 463-64.

51. See A. MASON, THE STATES RIGHTS DEBATE: ANTIFEDERALISM AND THE CONSTITUTION 49-52 (2d ed. 1972). Even after the Convention, Madison continued to defend a congressional veto on state legislation. See letter from Madison to Jefferson, October 24, 1787, in 12 THE PAPERS OF THOMAS JEFFERSON 274 (J. Boyd ed. 1955).

52. Roche, *The Founding Fathers: A Reform Caucus in Action,* in AMERICAN CONSTITUTIONAL LAW: HISTORICAL ESSAYS 12, 45 (L. Levy ed. 1966).

53. See, *e.g.,* the evidence reviewed in 3 I. BRANT, JAMES MADISON 101-05 (1950). See also MCDONALD, FORMATION 167-70.

the veto on state legislation, the delegates approved what would become the Supremacy Clause in Article VI of the final Constitution.[54] The Committee of Detail, meeting in late July and early August, declined to include a general grant of legislative power, but inserted the necessary and proper clause at the end of Congress's enumerated powers.[55] Another effort in late August to include a general grant of legislative power also failed, but the Convention now gave Congress "power to lay and collect taxes duties & excises, to pay the debts and provide for the common defence & general welfare."[56] Finally, the delegates accepted ratification by conventions drawn from the people, concluding this would give the new government a legal base independent of the state governments, just as Madison had explained to his correspondents prior to the convention.[57]

B. *The Problem of Sovereignty*

Besides approving novel nationalistic features which would soon require defense at the state level, the delegates were true to the concerns of the 1780s in puzzling over sovereignty. This occurred particularly in the course of describing the relationship of state and central governments within the new system.

In an early discussion of the Virginia plan, Madison recorded Gouverneur Morris as "explain[ing] the distinction between a *federal* and *national, supreme,* Govt.; the former being a mere compact resting on the good faith of the parties; the latter having a compleat and *compulsive* operation. He contended that in all communities there must be one supreme power, and one only."[58] Later, Hamilton was blunter: "The general power whatever be its form if it preserves itself, must swallow up the State powers.

54. 2 Farrand, Records 27-29 (Madison's notes, July 17).

55. Compare *id.* at 131-32 (proceedings referred to the Committee of Detail) with *id.* at 182 (Madison's notes of report of Committee of Detail, August 6). For speculation that the change occurred during discussion within the Committtee, see 3 Brant, *supra* note 53, at 117.

56. 2 Farrand, Records 367, 473 (Journal, August 22, 31), 497, 499 (Madison's notes, September 4). See 3 Brant, *supra* note 53, at 132-34, 139.

57. To trace this provision, see index entries under "Ratification" in 4 Farrand, Records 203-04; for the major debates, see 1 *id.* at 122-23; 2 *id.* at 88-94 (Madison's notes, June 5, July 23).

58. 1 *id.* at 34 (Madison's notes, May 30).

[O]therwise it will be swallowed by them. . . . Two Sovereignties can not co-exist within the same limits."[59] When William Johnson argued that the states had a dual nature—"in some respects the States are to be considered in their political capacity, and in others as districts of individual citizens"[60]—Madison denied it. "Some contend that states are sovereign," he charged, "when in fact they are only political societies. There is a gradation of power in all societies, from the lowest corporation to the highest sovereign. The states never possessed the essential rights of sovereignty. These were always vested in Congress."[61] Madison thereby sought to play down the contrast between the Articles of Confederation and the Virginia Plan. Yet in the process of doing so, he revealed his commitment to the premise of Morris, Hamilton, and, as later became apparent, Johnson: sovereignty was by nature indivisible, or very close to it.[62]

Especially telling with respect to the pervasiveness of this premise is an exchange between William Paterson and James Wilson. "A confederacy supposes sovereignty in the members composing it & sovereignty supposes equality," argued Paterson, adding: "If we are to be considered a nation, all State distinctions must be abolished, [and] the whole must be thrown into a hotchpot . . . ." Rejecting Paterson's conclusion, Wilson nevertheless adhered to the same understanding of sovereignty: "If N.J. will not part with her Sovereignty it is vain to talk of Govt." They agreed: the states individually might be sovereign, or the nation, but not both.[63]

Further insight comes from the debate over the treason clause.[64]

59. *Id.* at 287 (Madison's notes; June 18).

60. *Id.* at 461 (Madison's notes, June 29).

61. *Id.* at 471 (Yates's notes, June 29). Accord, *id.* at 463-64 (Madison's notes, June 29). See also *supra* n. 50 and accompanying text.

62. Cf. letter from Madison to Jefferson, October 24, 1787, *supra* note 51, at 273-74: "Without such a check in the whole over the parts, our system involves the evil of imperia in imperio. If a compleat supremacy some where is not necessary in every Society, a controuling power at least is so . . . ."

63. 1 Farrand, Records 178, 180 (Madison's notes, June 9). Paterson later repeated his comment; see *id.* at 251 (Madison's notes, June 16).

64. See 2 *id.* at 345-50 (Madison's notes, August 20). See also B. Chapin, The American Law of Treason: Revolutionary and Early National Origins *passim* and esp. 81-84 (1964); A. McLaughlin, A Constitutional History of the United States 178 & n. 36 (1935).

Here one concern was whether or not the wording of the clause should allow treason as a crime against individual states as well as against the United States. Treason being a breach of allegiance to the sovereign, the delegates were quite aware of the clause's implications for the question of the locus and nature of sovereignty. To the initial wording allowing treason against both individual states and the United States, William Johnson protested "that Treason could not be both agst. the U. States–and individual states; being an offense agst the Sovereignty which can be but one in the same community[.]"[65] The meaning of subsequent debate on the clause becomes clouded because of another concern: the possibility that double jeopardy would arise if states could punish treason.[66] Nonetheless, Johnson twice reiterated that the indivisibility of sovereignty meant treason could be committed only against the United States.[67] The record strongly hints Wilson took the same position.[68]

On the other hand, two members–George Mason of Virginia and Oliver Ellsworth of Connecticut–seemingly admitted the divisibility of sovereignty while discussing treason.[69] But it was also Ellsworth who explained what perhaps made acceptable the final wording which retained the possibility of state treasons. "There can," he noted, "be no danger to the Genl authority from this; as the laws of the U. States are to be paramount."[70]

These various exchanges suggest most of the delegates were highly conventional in their convictions about sovereignty's indivisibility. A few evidently disagreed, but failed to develop their views in much detail. In this regard, then, the Convention did not transcend existing conceptions. Significantly, however, the method adopted for the Constitution's ratification laid a basis for preserving the orthodox notion of indivisibility while departing from the Blackstonian view of sovereignty's locus.

Debate over the ratification process showed a substantial consensus for approval by the people.[71] Legislatures had no power to

65. 2 Farrand, Records 346.
66. See esp. Madison's comments, *id.* at 346, 349.
67. *Id.* at 347.
68. See *id.* at 348.
69. See *id.* at 347, 349.
70. *Id.* at 347.
71. See 1 *id.* at 122-23; 2 *id.* at 88-94 (Madison's notes, June 5, July 23).

ratify the Constitution, explained Mason. "They are the mere creatures of the State Constitutions, and cannot be greater than their creators."[72] Only if the new system rested on the people, Madison added, would it be a constitution rather than a treaty and thereby require judges to overturn conflicting state legislation.[73] And even those accepting legislative ratification argued on balance that practice had established this procedure as the equivalent of popular ratification.[74] In short, they too allowed that the new system needed a popular base.

One of those who defended legislative ratification in these terms alluded to a significant trend. Oliver Ellsworth "observed that a new sett of ideas seemed to have crept in since the articles of Confederation were established. Conventions of the people, or with power derived expressly from the people, were not then thought of."[75] By 1787, however, the people themselves had ratified state constitutions.[76] Moreover, opponents of particular policies within individual states had argued that state legislatures could not transgress established constitutional limits precisely because those limits reflected the *continually* binding command of the sovereign people.[77]

By 1787, that is, some Americans came to regard the people as holding, albeit in a hard-to-describe fashion, not only what can be called ultimate political sovereignty but also legal sovereignty. Blackstone's denial to the people of the latter, short of a return to the state of nature, would perhaps have made sense to Luther Martin, who "conceived . . . that the people of the States[,] having already vested powers in their respective Legislatures, could not resume them without a dissolution of their Governments."[78] But Martin's view did not prevail in Philadelphia. Some

72. *Id.* at 88.

73. *Id.* at 93.

74. See comments of Ellsworth and King, *id.* at 91, 92. Although willing to accept legislative ratification as the equivalent of ratification by conventions drawn from the people, King preferred the latter; see *id.* 92. But see Luther Martin's comments, *id.* at 89-90, 476.

75. 2 Farrand, Records 91.

76. In Massachusetts (1780) and New Hampshire (1784). See Federal and State Constitutions 956, 1280.

77. See Wood, Creation 369-89. See also *id.* at 430-67.

78. 1 Farrand, Records 341 (Madison's notes, June 20). Accord, *id.* at 437

there feared and others hoped the new government would legally overshadow or even absorb the states; but they did not follow Martin in holding that the act of ratification by the people of the states would *destroy* the state governments. Instead, they apparently concluded that by and through an appropriate expression, the people might control government on an ongoing basis, as well as establish or destroy it.[79]

As it finally emerged from Philadelphia, the Constitution contemplated a "general government partly federal and partly national," to borrow Ellsworth's characterization during the Convention.[80] It thereby lay open to the charge of dooming the states through embodying the solecism of *imperium in imperio.* Yet its provisions soon turned out to allow explanations, grounded in new insights, which denied both that any such solecism inhered in the document and that any further guarding of the states was necessary in order to preserve them.

IV. The Ratification Debates, 1787-1788

Once the contest shifted to the states, its terms changed greatly. In a real sense the measure of the nationalists' strength in the Federal Convention was William Paterson's New Jersey Plan. Although commonly contrasted to Randolph's nationalistic Virginia Plan, Paterson's proposal contained a supremacy clause similar to that of Article VI of the final document; it thereby anticipated a strong central government.[81] This meant the Convention could work for compromise within a framework of commitment to

---

(Madison's notes, June 27): ". . . to resort to the Citizens at large for their sanction to a new Governt. will be throwing them back into a State of Nature: . . . the dissolution of the State Govts. is involved in the nature of the process: . . . the people have no right to do this without the consent of those to whom they have delegated their power for State purposes; through their tongue only they can speak, through their ears, only, can hear . . . ." Cf. Martin's comments cited in note 74 *supra.*

79. See generally 3 Brant, *supra* note 53, at 155.

80. 1 Farrand, Records 474 (Yates's notes, June 29).

81. See *id.* at 242-45 (Madison's notes, June 15); McDonald, Formation 167-70; Diamond, *What the Framers Meant by Federalism,* in A Nation of States 25, 38 (R. Goldwin ed. 1974).

strengthened central authority. The hard core opposition comprised perhaps a half dozen delegates out of the fifty-five who attended at one time or another.[82] In the states, on the other hand, the stakes were all or nothing, at least for the moment. Because the Federalists contended conditional ratification was no ratification, the state conventions had to accept or reject the document outright. Moreover, although the Antifederalists suffered defeat in the end, they had substantial strength in key states and included some articulate spokesmen.[83]

Opposition in the states had important consequences. It forced the Federalists to innovation in their arguments which exceeded, at least in its explicitness, the inventiveness they had displayed in the Philadelphia Convention. Their arguments nevertheless failed to convince the opposition, so they turned to strategic innovation by promising subsequent amendments to the Constitution, including a reserved powers amendment. But, while insufficient in itself, the Federalists' argumentative inventiveness had the effects of contributing an increased acuity to contemporary understandings of the Constitution and the issues surrounding it, and thus also of affecting how Americans soon understood the Tenth Amendment.

A. *The Antifederalist Challenge*

Assessing the proposed Constitution, Samuel Adams wrote Richard Henry Lee: "I stumble at the Threshold. I meet with a National Government, instead of a Federal Union of Sovereign States."[84] In the Virginia convention, Patrick Henry charged: "The question turns . . . on that poor little thing–the expression, We, the *people,* instead of the *states,* of America." This phrasing had produced an "alarming transition, from a confederacy to a consolidated government."[85] The charge of consolidation

82. See McDONALD, FORMATION 187-88; 3 BRANT, *supra* note 53, at 154-60; Roche, *supra* note 52, *passim.*

83. See, *e.g.,* R. RUTLAND, THE ORDEAL OF THE CONSTITUTION: THE ANTIFEDERALISTS AND THE RATIFICATION STRUGGLE OF 1787-1788, *passim* [hereinafter cited as RUTLAND, ORDEAL]; MAIN, ANTIFEDERALISTS 119-248.

84. Quoted in RUTLAND, ORDEAL 81.

85. 3 THE DEBATES IN THE SEVERAL STATE CONVENTIONS ON THE ADOPTION OF THE FEDERAL CONSTITUTION 44 (J. Elliot ed., 2d ed. 1888) (June 5, 1788) [hereinafter cited as ELLIOT, DEBATES].

was, in fact, the most frequent Antifederalist objection to the Constitution.[86]

Many features of the document warranted the charge in Antifederalist eyes, but especially frightening were the undefined powers implied by terms like "necessary and proper," "general welfare," and "supreme law of the land." As Merrill Jensen has commented, nothing in the Articles of Confederation "remotely resembled such phrases."[87] The Constitution's enumeration meant little in the face of this unguarded wording.[88] One of Samuel Adams' correspondents was "more & more persuaded, that it is a Plan, that the common People can never understand–That if adopted–The Scribes & Pharisees only will be able to interpret, & give it a Meaning."[89] A Maine Antifederalist pursued the same thought: "There is a certain darkness, duplicity and studied ambiguity of expression running through the whole Constitution. . . . As it now stands but very few individuals do or ever will understand it, consequently Congress will be its own interpreter."[90] The object could only be the destruction of the states; and not only did the wording of the document point in that direction. Enough about the nationalists' original designs in Philadelphia had leaked out to give the charge additional credence.[91]

The Constitution met added difficulty because the Antifederalists by and large still viewed sovereignty in orthodox terms, stress-

86. For a sampling of other Antifederalist remarks to this end, see WOOD, CREATION, 526-27; MAIN, ANTIFEDERALISTS 120-26. For additional citations, see Ranney, *The Bases of American Federalism,* 3 WM. AND MARY Q. 1, 25 n. 97 (3rd ser. 1946).

87. MAIN, ARTICLES 242.

88. See MAIN, ANTIFEDERALISTS 122-23, 153-55; THE ANTIFEDERALISTS lxii-lxxiv (C. Kenyon ed. 1966) [hereinafter cited as KENYON, ANTIFEDERALISTS]. Hamilton commented that the necessary and proper clause and the supremacy clause "have been the sources of much virulent invective and petulant declamation against the proposed constitution, they have been held up to the people, in all the exaggerated colours of misrepresentation, as the pernicious engines by which their local governments were to be destroyed and their liberties exterminated . . . ." THE FEDERALIST, No. 33 at 204. For one of the best developed Antifederalist statements, see Robert Yates, *The Letters of Brutus,* 1788, excerpted in MASON, *supra* note 51, at 107-13.

89. Quoted in RUTLAND, ORDEAL 96-97.

90. Quoted in MAIN, ANTIFEDERALISTS 153.

91. See, *e.g.,* Luther Martin, *Genuine Information, supra* note 45.

ing not only its supreme, uncontrolled, and indivisible nature, but also its locus in the tangible organs of government, especially the legislature with its power of taxation.[92] "As to direct taxation—give up this, and you give up everything," warned William Grayson in Virginia, "as it is the highest act of sovereignty." "How are two legislatures to coincide," he asked, "with powers transcendent, supreme, and omnipotent? for such is the definition of a legislature."[93] In a careful and widely circulated analysis of the Constitution, the minority members of the Pennsylvania convention gave the objection full expression:

> We apprehend that two coordinate sovereignties would be a solecism in politics. That therefore as there is no line of distinction drawn between the general and state governments; as the sphere of their jurisdiction is undefined, it would be contrary to the nature of things, that both should exist together, one or the other would necessarily triumph in the fullness of dominion. However, the contest could not be of long continuance as the state governments are divested of every means of defense, and will be obliged by 'thc supreme law of the land' *to yield at discretion.* . . . [Moreover, the] legislative power vested in Congress by the foregoing recited sections is so unlimited in its nature; may be so comprehensive, and boundless [in] its exercise, that this alone would be amply sufficient to annihilate the state governments, and swallow them up in the grand vortex of general empire.[94]

As Gordon Wood has summarized the Antifederalist attack, the doctrine of legislative sovereignty which had so influenced the coming of war with Britain "was now relentlessly thrown back at the Federalists by the opponents of the Constitution. . . . The

92. See Main, Antifederalists 123-24; Wood, Creation 526-29. Cf. Rutland, Ordeal 56-57, 81, 202, 212. See also Professor Corwin's discussion of Luther Martin's assimilating "state" and "government" in Corwin, *National Power and State Interposition, 1787-1861,* 3 Selected Essays on Constitutional Law 1171, 1175 (Assoc. Am L. Schools ed. 1938).

93. 3 Elliot, Debates 280-81 (June 11, 1788). Accord, 2 *id.* at 403 (Thomas Tredwell, N.Y. convention, July 1, 1788).

94. 2 The Documentary History of the Ratification of the Constitution (*Ratification of the Constitution by the States: Pennsylvania*) 628-29 (M. Jensen ed. 1976) [hereinafter cited as *Pa. Ratification Debates*].

logic of the doctrine of sovereignty required either the state legislatures or the national Congress to predominate."[95]

B. *The Federalist Response*

Federalist responses to charges of consolidation and congressional sovereignty sometimes skirted the issues, denied the problems, or charged the Antifederalists themselves with inconsistency. Madison, for example, held that the American Revolution had been fought to secure the happiness of the American people, not to guarantee state sovereignty.[96] He and Hamilton explained that, contrary to the Antifederalist fears, the attachment of citizens to their local governments would jeopardize the central government.[97] Hamilton ridiculed the Antifederalists' inconsistency in conceding the need for a strengthened federal union, yet at the same time demanding a "complete independence in the members." It was they who "seem[ed] to cherish with blind devotion the political monster of an *imperium in imperio*."[98]

Federalists met the attack directly, as well. Because the Constitution carefully enumerated the powers of Congress, no further safeguards were necessary. Unlike the state governments, the new central government could exercise only those powers delegated it. However, explaining this view led the Federalists off in various directions.

1. Divisible Sovereignty. After 1789, the *Federalist* papers of "Publius" quickly became the authoritative exposition of the new system. In fact, of course, the Constitution's proponents produced numerous other tracts, together exceeding the *Federalist* in volume, and this literature probably offers a better guide to how most Federalists viewed the Constitution *during* the ratification fight. Reviewing this material, Herbert Storing found the Constitution's advocates "usually . . . conceded the historical and legal priority of the states."[99] Roger Sherman of Connecticut offered what Storing calls a "rather typical description": "The Powers vested in

95. Wood, Creation 527.

96. The Federalist No. 45 at 309.

97. *Id.*, Nos. 17 (Hamilton) and 45 (Madison).

98. *Id.*, No. 15 at 93.

99. Storing, *The "Other" Federalist Papers: A Preliminary Sketch*, 6 Pol. Sci. Reviewer 215, 220 (1976).

the federal government are clearly defined, so that each state will retain its sovereignty in what concerns its own internal government, and a right to exercise every power of a sovereign state not particularly delegated to the government of the United States."[100]

Sherman's view implies the *divisibility* of sovereignty. Certain language in the *Federalist* itself arguably carries the same implication. "[A]s the plan of the Convention aims only at a partial union or consolidation," wrote Hamilton, "the State Governments would clearly retain all the rights of sovereignty which they had and which were not by that act *exclusively* delegated to the United States."[101] Madison claimed that "the States will retain . . . a very extensive portion of active sovereignty,"[102] and that state equality in the Senate "is at once a constitutional recognition of the portion of sovereignty remaining in the individual states, and an instrument for preserving that residuary sovereignty."[103]

What to make of language of this sort is problematic. Almost a half century later in his *Commentaries on the Constitution of the United States,* Joseph Story carefully and clearly distinguished sovereignty in its ultimate sense–that is, as the source of, and final control on, government–from the *powers* of sovereignty.[104] Because sovereignty in either of these senses may loosely be labeled "sovereignty," not all language suggesting a division of sovereignty necessarily indicates a division of sovereignty in its ultimate sense. Yet Sherman's remark seems clearly to refer to a division of sovereignty in this ultimate sense, not to a division between different governments of different *powers* of sovereignty. Anyone viewing sovereignty as that power which was supreme and uncontrolled, *and* which inhered *in government,* would likely also interpret the quoted comments from Madison and Hamilton as attempting a division of what was by nature indivisible. In short, these Federalist

100. Quoted in *id.* at 222.
101. THE FEDERALIST No. 32 at 200.
102. *Id.,* No. 45 at 310.
103. *Id.,* No. 62 at 417. The conclusion that sovereignty is divisible can also be read into the letter from Gen. Washington, September 17, 1787, officially transmitting the Constitution to the President of Congress; see 2 FARRAND, RECORDS 666.
104. See 1 J. STORY, COMMENTARIES ON THE CONSTITUTION OF THE UNITED STATES 191-94 (secs. 207-08) (1833).

defenses against charges of consolidation probably carried little weight with the unconverted, whatever their authors themselves meant by them.

2. A Changed Locus of Sovereignty. A few Federalists, most notably Wilson in Pennsylvania and Madison elsewhere in the *Federalist* and the Virginia convention, offered more careful explanations. These differed in complexity, however, and to some extent in content.

In a sense Wilson began his defense in Philadelphia. During the debate over equal representation in the Senate he had explained: "The Genl. Govt. is not an assemblage of States, but of individuals for certain political purposes–it is not meant for the States but for the individuals composing them . . . ."[105] "The truth is," he elaborated in the Pennsylvania convention in November 1787, "that, in our governments, the supreme, absolute, and uncontrollable power *remains* in the people." The central government was "founded on a representation of the *whole* Union; whereas the government of any particular state [was] founded only on a representation of the part . . . ."[106] Or, as he put it later, "the power[s] both of the general government, and the state governments, under this system are acknowledged to be so many emanations of power from the people."[107] As for the charge that the Constitution transferred sovereignty from the state governments to the central government, "the sovereignty [was] not in the possession of the state governments, therefore it cannot be transferred from these to the general government."[108] The Constitution accordingly neither divided the indivisible nor transferred it.

A practical problem remained. Easy as it was "to discover a proper and satisfactory principle on the subject," Wilson admitted that in dividing the *powers* of sovereignty, to use Story's later terminology, the line between the state and central authorities could not be drawn with "mathematical precision." Yet he flattered himself "that upon the strictest investigation, the enumeration will be found to be safe and unexceptionable; and accurate too in as great

105. 1 FARRAND, RECORDS 406 (Madison's notes, June 25, 1787).

106. *Pa. Ratification Debates* 361, 355-56. Accord, id. at 471-74, 555-56.

107. *Id.* at 559.

108. *Id.* at 560.

a degree as accuracy can be expected in a subject of this nature."[109]

This "safe and unexceptionable" enumeration posed no threat to the state governments; indeed, the system was "founded on their existence."[110] But whatever role the existing states played in forming the new government, Wilson's defense of it stressed the sovereignty of the people of the *United* States and, concomitantly, an attack on state sovereignty. Fittingly, his last major address to the Pennsylvania convention proclaimed that "by adopting this system, we become a NATION . . . ."[111]

Far more complex, Madison's defense came close to denying that a single principle informed the relations between the state and general governments under the Constitution. Certainly no principle could be found *in* the Constitution.[112] Concerning the different bases of representation in the Senate and House of Representatives, he advised that "[i]t is superfluous to try by standards of theory a part of the constitution which is allowed on all hands to be the result not of theory, but 'of a spirit of amity, and that mutual deference and concession which the peculiarity of our political situation rendered indispensable.' "[113]

When he sought to describe the Constitution, Madison elaborated a characterization advanced earlier by others in the Philadelphia Convention.[114] Defining the *"federal* form . . . as a *confederacy* of sovereign states" and a *"national* government . . . as a *consolidation* of the States,"[115] he found that "[t]he proposed

109. *Id.* at 355, 496.

110. *Id.* at 560. Wilson did not here explain how the general government was founded on the existence of the states, but he had earlier described the role of the state legislatures in electing the new Senate. See *id.* at 170.

111. *Id.* at 581.

112. Note should be taken of Professor Jaffa's explication of *Federalist 43* (see Jaffa, supra *note* 34 at 136-37), in which he shows that Madison held the central government's authority rested ultimately on the *extra*-constitutional principle of self-preservation.

113. THE FEDERALIST, No. 62 at 416.

114. 1 FARRAND, RECORDS 331 (King's notes, June 19), 468 (Madison's notes, June 29).

115. THE FEDERALIST No. 39 at 253. See the comparable definition offered by Gouverneur Morris in Philadelphia, 1 FARRAND, RECORDS 34. Madison defined the terms in this way in the course of summarizing the Antifederalist case against the Constitution, but his definitions have general applicability for the 1780s. See Diamond, *supra* note 81 at 27-28.

Constitution . . . [was] in strictness neither a national nor a federal constitution; but a composition of both." Giving the system a national character were the House of Representatives and the operation of the government directly on individuals; giving it a federal character were the Senate and the limitation on the government's operation "to certain enumerated objects only."[116]

In these regards, the complexity of Madison's defense as compared to Wilson's begins to appear, but the differences between the two became more pronounced. Describing the source of the central government's power, Madison did not claim–and could not, in view of the Constitution itself–that the Constitution rested on acts by the states as sovereign units in the sense that the Antifederalists regarded them as sovereign. The new system, that is, did not rest on acts of the existing legislatures of the states. Instead, it rested "on the assent and ratification of the people of America," but not, as Wilson would have it, on the people of the United States collectively. Rather, its source was the people within the individual states acting as the supreme power within each state. Also, the new system would go into operation not by a majority vote of the people of America as a whole, nor upon the vote of a majority of the states, but only upon "the *unanimous* assent of the several States that are parties to it . . . ."[117] And regarding the amendment process, the system was "neither wholly *national,* nor wholly *federal.*" A majority of the people could not amend it by virtue of being a majority; voting on amendments was by states; yet a state could be bound against its will.[118]

Neither Wilson's nor Madison's elaboration convinced the Antifederalists. Wilson denied any theoretical base for state sovereignty. Madison omitted quite such a blatantly heretical discussion of sovereignty, but was just as unpersuasive. Patrick Henry, leading the Virginia opposition, offered ridicule in reply to Madison: "We may be amused, if we please, by a treatise of political anatomy. In the brain it is national; the stamina are federal; some limbs are federal, others national." He was not fooled by this

116. The Federalist No. 39 at 254-57. Accord, 3 Elliot, Debates 93-95 (Virginia convention, June 6, 1788).

117. The Federalist No. 39 at 254.

118. *Id.* at 257.

"most curious anatomical description," for "[t]o all common purposes of legislation, it is a great consolidation of government."[119] Nor should Henry's conclusion be attributed too readily to perversity (not that Henry was incapable of it). Seeing sovereignty as indivisible, John Adams, too, thought that the new government was "wholly national," for to conclude otherwise ran head on into what he saw as the "solecism" of "imperium in imperio."[120]

C. *Demands for a Reserved Powers Clause*

Whether emphasizing a division of sovereignty between the states and the proposed general government, or the absolute sovereignty of the American people, or the partly federal-partly national character of the Constitution, the Federalists concluded that the actual powers of the new government would be limited to those enumerated in the Constitution. Unconvinced by any of these explanations, the Antifederalists were likewise unconvinced by the Federalists' accompanying denials of the need for a bill of rights, in favor either of individuals or of the states, and demands for one quickly surfaced.

1. Background in Philadelphia. Earlier, the subject of explicitly limiting central power over the states had arisen on at least four occasions during the Philadelphia convention. The first proposal came in Charles Pinckney's plan of government, introduced on May 29, the same day as the Virginia Plan. This provided: "Each State retains its rights not expressly delegated."[121] So far as the records indicate, the restriction received no further hearing. The second occasion came a month later while the basis of representation in the Senate was before the Convention. Rufus King unsuccessfully urged including a statement of "the Rights of States" in the Constitution, but the record is vague regarding what he intended.[122]

As the Convention was about to adjourn, two further sugges-

119. 3 Elliot, Debates 171 (June 9, 1788). See letter of Luther Martin, *Maryland Journal,* March 21, 1788, in Essays on the Constitution of the United States . . . 1787-1788 at 360, 366-68 (P. Ford ed. 1892).

120. Quoted in Wood, Creation 580-81

121. 3 Farrand, Records 607 (italics omitted) (Farrand's reconstruction of Pinckney's plan).

122. See 1 *id.* at 493 (Madison's notes, June 30).

tions appeared. One came on September 12 when the issue arose of including a guarantee of jury trials in civil cases. Rather than writing an overly specific guarantee into the Constitution itself, George Mason argued that "[a] general principle laid down on this and some other points would be sufficient." Prefacing the Constitution with a bill of rights "would give a great quiet to the people," and by consulting the existing state constitutions one could quickly be drawn. When Roger Sherman protested that the existing state declarations provided adequate guarantees and would remain in force under the Constitution, Mason answered that laws of the general government would be paramount over the states. His pleas unavailing, Mason's motion was unanimously rejected.[123]

Precisely what rights the Convention would have included, had it accepted Mason's motion, is speculative but worth consideration. Mason's own involvement in drafting Virginia's wide ranging declaration of rights in 1776 hints *he* minimally had in mind a guarantee of individual rights.[124] But judging from his remarks on September 12, he also recognized the need for a limitation on central power to override state guarantees. Moreover, by 1787 each of six states had some kind of general reservation of power in favor of the people or the state, either in the state constitution or in a separate bill of rights. Dating from 1776, the Pennsylvania declaration provided that the people of the state had "the sole, exclusive and inherent right of governing and regulating the internal police of the same."[125] The Maryland and North Carolina declarations of the same year contained sections practically identical to Pennsylvania's.[126] The New York constitution of 1777 forbade exercise of any authority over the people "but such as shall be derived from and granted by them."[127] Showing the influence of Article II of the Confederation, the Massachusetts bill of rights of 1780, which New Hampshire copied almost verbatim in 1784, asserted: "The people of this commonwealth have the sole and

123. 2 *id.* at 587-88 (Madison's notes, September 12).

124. See Federal and State Constitutions 1908-09; R. Rutland, The Birth of the Bill of Rights, 1776-1791 at 32-40 (1955) [hereinafter cited as Rutland, Bill of Rights]. See also *id.* at 106-25.

125. Federal and State Constitutions 1540.

126. See *id.* at 817, 1409.

127. *Id.* at 1332.

exclusive right of governing themselves, as a free, sovereign, and independent State, and do, and forever hereafter shall, exercise and enjoy every power, jurisdiction, and right, which is not, or may not hereafter be, by them expressly delegated to the United States of America, in Congress assembled."[128]

These state provisions, plus the existence of Article II, suggest that in drafting a bill of rights the Convention would have included a reserved powers clause. But the conclusion is less certain in view of what happened three days later when Sherman, despite his earlier reply to Mason, made the fourth suggestion for an explicit limitation in favor of the states. With Article V describing the amendment process under review, Sherman "expressed his fears that three fourths of the States might be brought to do things fatal to particular States . . . ." He therefore moved that under Article V "no State shall without its consent be affected in its internal police, or deprived of its equal suffrage in the Senate."[129] The motion failed on an eight-to-three vote,[130] yet Gouverneur Morris's subsequent motion simply to forbid any deprivation of state equality in the Senate passed without opposition.[131] Perhaps, consistent with the later Federalist defenses of the Constitution, the framers in Philadelphia indeed thought a reserved powers clause either unnecessary or perhaps even unwise. Speculation aside, omission of such a clause had a clear impact on the coming ratification struggle.

2. *Maneuvering in the States.* Debate in Pennsylvania, the first state to hold its convention, anticipated much of what followed.[132] Even before the state's convention met, Samuel Bryan published a widely reprinted tract scoring the Constitution for its lack of a guarantee of liberty like the one in Pennsylvania's own constitution. The next day, October 6, James Wilson replied in a speech which similarly received wide circulation and contained what became the standard Federalist explanation for the absence of a bill of rights.[133]

128. *Id.* at 958, 1281.
129. 2 Farrand, Records 629-30 (Madison's notes, September 15).
130. *Id.* at 630.
131. *Id.* at 631.
132. See generally Rutland, Bill of Rights 126-89.
133. See *Pa. Ratification Debates* 128, 158-59, 166, 167-72.

The example of state declarations of rights was irrelevant, Wilson explained, for the state legislatures had "every power and authority" which the people had not explicitly reserved; "[b]ut in delegating federal powers, another criterion was necessarily introduced, and the congressional authority [was] to be collected, not from tacit implication, but from the positive grant expressed in the instrument of union." It was evident, he said, that with the states "everything which is not reserved is given," but in the new Constitution "the reverse of the proposition prevails, and everything which is not given is reserved." Thus the new government had no power over the press, for example, so "it would have been merely nugatory to have introduced a formal declaration upon the subject"; it might even have implied that some authority did exist in the area.[134] Further, as he later developed the point, many powers and rights could not be "particularly enumerated," yet "an imperfect enumeration would throw all implied power into the hands of the government; and the rights of the people would be rendered incomplete."[135]

The Pennsylvania Antifederalists remained unconvinced. Robert Whitehill observed that the Constitution in fact contained a rudimentary bill of rights in its provisions regarding habeas corpus and jury trial in criminal cases. This, by Wilson's own logic, threatened the great body of *other* rights. Because the Constitution's consolidating features would destroy "the powers and sovereignty of the several states," the people would ultimately have no protection against their rulers.[136]

This line of attack became common as the ratification process continued, but it proved no match in the Pennsylvania convention for the Federalists' two-to-one voting edge. Accordingly, the Antifederalists failed in their last minute effort to postpone action in order to consider amendments to the Constitution.[137] Significantly, though, the amendments offered by Whitehill included a close paraphrase of the Confederation's Article II, and this was re-

134. *Id.* at 167-68.
135. *Id.* at 388.
136. *Id.* at 397-98, 427.
137. See MAIN, ANTIFEDERALISTS 288; RUTLAND, ORDEAL 55-58; *Pa. Ratification Debates* 589-91, 597-600.

peated in the separate statement subsequently issued by the Pennsylvania minority. The omission of such a guarantee, charged the dissenters, fitted in with "the plan of consolidation" underlying the Constitution.[138]

But defeat of amendments in Pennsylvania marked only the beginning. Failing otherwise to develop a unified critique of the Constitution, the Antifederalists agreed on the need for a bill of rights,[139] and their insistence produced results. As elaborated, for example, by Wilson in Pennsylvania, Hamilton in *Federalist 84,* and Madison in the Virginia convention,[140] the Federalist argument that the people needed no guarantee of rights against the sovereign, because the people themselves were sovereign, could not overcome widespread suspicion of government. An obvious corollary to the notion of state legislative sovereignty, the old view persisted of a constitution as a "compact between the Governors and the Governed."[141] At a less theoretical level, the locus of sovereignty made little difference, for the new government would hold broad powers anyway, and these, if unguarded, might serve as bases for attacks on individuals and states under the necessary and proper clause. Then, too, the existence of a few guarantees raised questions about whether the Constitution compromised other rights. Finally, the supremacy clause invalidated state guarantees.[142] "The victory in the argument clearly lay with the Antifederalists," concludes Cecilia Kenyon, "for the reasoning of their opponents was

138. *Id.* at 599, 624, 629

139. See RUTLAND, ORDEAL 313; KENYON, ANTIFEDERALISTS, editor's introduction, *passim.*

140. See text accompanying notes 134-35 *supra;* THE FEDERALIST No. 84 at 578; 3 ELLIOT, DEBATES 620 (June 24, 1788).

141. *Letters of "John DeWitt,"* in KENYON, ANTIFEDERALISTS 98. See WOOD, CREATION 540-43.

142. See KENYON, ANTIFEDERALISTS lxx-lxxi; RUTLAND, ORDEAL *passim;* and esp. the common sense view of the "Federal Farmer": "The truth is, . . . it is mere matter of opinion, and men usually take either side of the argument, as will best answer their purposes: But the general presumption being, that men who govern, will in doubtful cases, construe laws and constitutions most favourably for increasing their own powers; all wise and prudent people, in forming constitutions, have drawn the line, and carefully described the powers parted with and the powers reserved." . . . *Letters from the Federal Farmer to the Republican,* 1787, in PAMPHLETS ON THE CONSTITUTION OF THE UNITED STATES . . . 1787-1788 at 277, 313 (P. Ford ed. 1888).

muddled, inconsistent, contradictory, and thoroughly unpersuasive."[143]

Convening on January 9, 1788, the Massachusetts convention proved pivotal. Five states had already ratified, but the Massachusetts Antifederalists held an initial majority of perhaps twenty out of the 310 delegates.[144] By late January the issue appeared in such doubt to the Federalist leaders that they developed a new tactic. Rufus King wrote to Madison: "We are now thinking of amendments to be submitted not as a condition of our assent & Ratification, but as the opinion of the Convention subjoined to Ratification."[145] Playing on Governor John Hancock's vanity with the prospect of the Vice-Presidency–there was even a chance at the Presidency if Washington declined or Virginia did not join the Union–the Federalists persuaded him to introduce recommendatory amendments. With these, the state's ratification would be unconditional, but Antifederalist fears could still be quelled. The amendments passed and the tactic succeeded: the final majority for ratification was 187 to 168.[146]

Massachusetts set the pattern for the remaining states. Of the additional five which ratified before the new government went into operation,[147] four adopted recommendatory amendments, including geographically vital New York and Virginia. In the fifth, Maryland, recommendatory amendments at least gained committee support.[148] Federalist acceptance of recommendatory amendments, was, says Robert Rutland, "the master stroke of the ratification campaign."[149]

3. The Problem of "Expressly." The sets of amendments offered by Massachusetts, New Hampshire, New York, South Carolina, and Virginia each contained a provision reserving to the states those powers not delegated to the central government; and

143. KENYON, ANTIFEDERALISTS lxxi.

144. MAIN, ANTIFEDERALISTS 203, 288.

145. Quoted in RUTLAND, ORDEAL 105.

146. See *id.* at 104-09; MAIN, ANTIFEDERALISTS 288.

147. North Carolina and Rhode Island failed to ratify.

148. For a summary of this sequence of events and of the amendments proposed, see E. DUMBAULD, THE BILL OF RIGHTS AND WHAT IT MEANS TODAY 14-30 (1957).

149. RUTLAND, ORDEAL 301. See *id.* at 33, 160-62, 197.

all but Virginia's would have reserved all powers not "expressly" or, as New York put it, "clearly" delegated.[150] The significance of such wording was well recognized. In North Carolina, which initially declined to ratify, Antifederalist Samuel Spencer held that a clause like the Confederation's Article II "would render a bill of rights unnecessary."[151] In the same vein in Massachusetts, Samuel Adams, now maneuvered into supporting the Constitution,[152] praised the reservation of all powers not expressly delegated as a summary of a bill of rights.[153] Federalist Charles Jarvis agreed that "by positively securing what is not expressly delegated, it leaves nothing to the uncertainty of conjecture, or to the refinements of implication, but is an explicit reservation of every right and privilege which is nearest and most agreeable to the people."[154] Their descriptions of the difficulties Article II caused the Confederation government disclose other Federalists knew the importance of "expressly." Scouting Patrick Henry's proposal in Virginia for this kind of restrictive amendment,[155] Governor Randolph provided an example when he asked if "these gentlemen [are] zealous friends to the Union, who profess to be so here, and yet insist on a repetition of measures which have been found destructive of it?"[156]

150. See 1 ELLIOT, DEBATES 322 (Massachusetts), 325 (South Carolina), 326 (New Hampshire), 327 (New York); 3 *id.* at 659 (Virginia). The unofficial Maryland and Pennsylvania recommendations also included reserved powers proposals using "expressly." See 2 *id.* at 550 (Maryland); *Pa. Ratification Debates* 624. Strictly speaking, the New York and South Carolina reserved powers proposals constituted each convention's "understanding" of the Constitution, but were put forward in the context of recommendations for amendment. In addition to its recommendation for a reserved powers amendment, Virginia included a reserved powers "understanding." See 3 ELLIOT, DEBATES 656.

151. 4 *id.* at 163 (July 29, 1788).

152. See MCDONALD, FORMATION 216 for the details of the maneuver.

153. 2 ELLIOT, DEBATES 131 (February 1, 1788).

154. *Id.* at 153 (February 2, 1788).

155. Elliot does not include the text of Henry's amendments, but Randolph's remarks, Henry's stated views, and other evidence, when taken together, indicate that Henry included "expressly" in his amendments. See 3 *id.* at 445-46, 593, 600-601, 622-23; THE BILL OF RIGHTS: A DOCUMENTARY HISTORY 1118 (B. Schwartz ed. 1971) (*Annals of Congress,* August 28, 1789) [hereinafter cited as DOC. HIST. OF BILL OF RTS.]

156. 3 ELLIOT, DEBATES 601 (June 24, 1788). See, *e.g.,* THE FEDERALIST No. 21 at 129-30, No. 44 at 302-05, for other Federalist critiques of the inclusion of "expressly" in the Articles.

In these ways, the ratification controversy indicates a contemporary understanding that absent "expressly" or "clearly," a reserved powers clause would have little meaning. There is more direct evidence, too. By the time George Wythe introduced into the Virginia convention a resolution of ratification which provided, among other things, that "every power not granted remains with the people,"[157] Madison had already attacked the inclusion of "expressly" in the Articles of Confederation.[158] He accordingly found unexceptionable Wythe's formulation omitting the term. "There cannot be a more positive and unequivocal declaration of the principle of the adoption—that every thing not granted is reserved. This is obviously and self-evidently the case, without the declaration."[159] Despite the importance assigned "expressly" by Henry and George Mason,[160] the Virginia convention deleted the term from its recommendatory amendments.[161]

The New York debate showed the same awareness. When the convention there turned to the necessary and proper clause, Robert Lansing introduced an amendment "[t]hat no power shall be exercised by Congress, but such as is expressly given by this Constitution; and all others, not expressly given, shall be reserved to the respective states, to be by them exercised."[162] Sometime before the state's final ratification, this was changed from a recommendatory amendment to an explanation "that every power, jurisdiction and right which is not by the said Constitution *clearly* delegated to the Congress of the United States, or the departments of government thereof, remains to the people of the several states, or to their respective state governments, to whom they may have delegated the same . . . ."[163] When the change occurred is uncertain,[164] but Gilbert Livingston's sketchy notes offer a clue to why

157. 3 ELLIOT, DEBATES 587, 653 (June 24, 1788).

158. See THE FEDERALIST No. 44 at 302-05.

159. 3 ELLIOTT, DEBATES 620 (June 24, 1788).

160. See *id.* at 441-42, 445-46, 622-23 (June 16, 24, 1788).

161. Elliot does not include the actual decision to delete, but Madison subsequently referred to it in Congress in 1789, and his recollection is consistent with other evidence herein reviewed. See citations in note 155 *supra.*

162. 2 ELLIOT, DEBATES 406 (July 2, 1788).

163. 1 *id.* at 327 (italics added).

164. From the material discussed in the text accompanying notes 165-66 *infra,* Professor Schwartz concludes the change occurred on July 19, 1788 (see DOC. HIST. OF BILL OF RTS. 854), but as I read it, the evidence is not as clear.

it was made. For the debate over the amendment, which still contained "expressly," Livingston recorded:

> explanitory amendt . . .
> 2d–No power to be exercised–but what is expressly given.
> Ham[ilton]–combats the propriety of the word "expressly"–congress one [are?] to regulate trade–now they must do a thousand things not expressly given–Virginia say not given
> Harper–the Committee may correct this.
> Yates–agrees–that in grantg genl powers–the powers to execute are implied.
> questn on the paragraph–agreed.[165]

That the change to "clearly" occurred at this time seems unlikely, for shortly afterwards Hamilton again mentioned the difficulty that "expressly" posed.[166] But the exchange once more indicates an understanding of the term's importance. (It reveals, too, that Virginia's omission of it drew notice.) By subsequently substituting "clearly" for "expressly," the convention likely intended to lessen the restrictive force of its "explanation."[167]

V. Framing the Tenth Amendment

Although endorsing Virginia's recommendatory amendments, Madison emerged from his state's convention at best as a lukewarm supporter of a bill of rights. He conceded, though, that a declaration of "political truths . . . in that solemn manner" could help "counteract the impulses of interest and passion," as well as provide the basis "for an appeal to the sense of the community" against the occasional usurpations by government.[168]

Once a member of Congress, Madison overcame his misgivings and quickly responded to President Washington's inaugural re-

165. Gilbert Livingston Papers, New York Public Library, July 19, 1788, in Doc. Hist. of Bill of Rts. 898.

166. *Id.* at 899.

167. See text accompanying note 192 *infra.*

168. Letter from Madison to Jefferson, October 17, 1787, excerpted in Mason, *supra* note 51, at 176, 177. See *id.* at 176-78; A. Koch, Jefferson and Madison: The Great Collaboration 52 (1964). For this entire section (*i.e.*, text accompanying notes 168-90), see generally Rutland, Bill of Rights 190-215.

minder to consider amendments.[169] Correspondence with Jefferson may have helped him appreciate the barrier a bill of rights would provide against usurpation.[170] Then, too, in five states a Federalist commitment to seek consideration of amendments had accompanied ratification and was probably crucial to it.[171] In Virginia itself, Madison's promise to support amendments helped him win a seat in the new House of Representatives.[172] On the ominous side, although ratifying unconditionally, New York had circularized the other states with a call for a second convention to consider revisions to the Constitution.[173] Sensing the diverse dangers such a convention would pose after the open airing of dissent in the state ratifying conventions, and perhaps overestimating support for a second convention, Madison saw the New York move as an added reason to act quickly in the new Congress on specific and limited amendments.[174]

In spite of opposition—other business should take priority, said the footdraggers[175]—Madison introduced amendments on June 8, 1789. His proposal for what became the core of the Tenth Amendment read: "The powers not delegated by this constitution, nor prohibited by it to the States, are reserved to the States respectively."[176] Attempts to add "or to the people" came on August 18

169. 1 MESSAGES AND PAPERS OF THE PRESIDENTS 45 (J. Richardson, comp. 1897) (April 30, 1789); DOC HIST. OF BILL OF RTS. 1012 (*Annals of Cong.*, May 4, 1789).

170. See letters from Jefferson to Madison, December 20, 1787, and March 15, 1789, excerpted in MASON, *supra* note 51, at 170-72, 180-82. See also KOCH, *supra* note 168, at 51-59.

171. See text accompanying notes 139-49 *supra*.

172. See 3 BRANT, *supra* note 53, at 240-42; RUTLAND, ORDEAL 297; letter from Madison to George Eve, January 2, 1789, in DOC. HIST. OF BILL OF RTS. 996-97.

173. Letter from George Clinton to the Several Governors, July 28, 1788, in 2 ELLIOT, DEBATES 413-14.

174. See letters from Madison to Jefferson, April 22, September 21, December 8, 1788, excerpted in MASON, *supra* note 51, at 173-74, 175-76, 178-79; RUTLAND, ORDEAL 279-300; 3 BRANT, *supra* note 53, at 264.

175. See DOC. HIST. OF BILL OF RTS, 1017-23 (*Annals of Cong.*, June 8, 1789).

176. *Id.* at 1028 (*Annals of Cong.*, June 8, 1789). Madison initially proposed incorporating his amendments in the text of the Constitution. The House eventually voted instead to append the amendments as a separate and supplementary section, some members arguing that this form was more convenient, and others that it was inappropriate to use the amendment process of Article V to change the body of a document which had been approved by conventions drawn from the people. See *id.* at 1066-75, 1126 (*Annals of Cong.*, August 13, 19, 1789).

and August 21. The latter move may have succeeded. The *Annals of Congress,* based on contemporary newspaper reports of the debates, shows it did, but the official journal neither reported the action nor included the phrase in its version of the amendment.[177] In any event, the Senate approved the Amendment in its final form, including "or to the people," on September 7, and a conference of the two Houses accepted it as the Twelfth Amendment on September 24.[178] The first two amendments failed to receive the required state ratifications, so the reserved powers amendment became the Tenth Amendment to the Constitution with Virginia's approval in December 1791 and the Secretary of State's official proclamation on March 1, 1792.[179]

Congressional debate on the Amendment allows fairly firm conclusions about its meaning to its framers. In his introductory speech, Madison explained that he found, "from looking into the amendments proposed by the State conventions, that several are particularly anxious that it should be declared in the constitution, that the powers not therein delegated should be reserved to the several States."[180] Yet, as already discussed, three of the states (or five, counting the unofficial Pennsylvania and Maryland recommendations) had included "expressly" in their proposals; one had included "clearly"; and only one–Madison's own Virginia–had recommended an amendment in the form Madison adopted, without either "expressly" or "clearly" to strengthen its restrictive force. Accordingly, if he meant to suggest that his amendment should entirely calm state anxieties, then he was misleading; but it seems unlikely that many were misled. His next comment was revealing: "Perhaps words which may define this [reservation of power] more precisely than the whole of the instrument now does, may be considered superfluous. I admit that they may be deemed unnecessary; but there can be no harm in making such a declaration, if the gentlemen will allow that fact is as stated. I am sure I

177. See *id.* at 1118, 1127-28 (*Annals of Cong.,* August 18, 21, 1789), 1124-25 (*House Journal,* August 21, 1789).

178. See *id.* at 1150-51, 1164, 1165 (*Senate Journal,* September 7, 24, 26, 1789) 1160-61 (*House Journal,* September 24, 1789).

179. See documents and editor's notes in *id.* at 1201-03.

180. *Id.* at 1033 (*Annals of Cong.,* June 8, 1789).

understand it so, and I do therefore propose it."[181] In sum, Madison viewed the Amendment as adding nothing to the existing Constitution.

Subsequent unsuccessful attempts to insert "expressly" reveal that Madison's colleagues shared his understanding. The first was on a motion by Thomas Tucker of South Carolina on August 18. Madison "objected to this amendment, because it was impossible to confine a Government to the exercise of express powers; there must necessarily be admitted powers by implication, unless the constitution descended to recount every minutia." Roger Sherman agreed with Madison, but Tucker protested he was being misunderstood. He "did not view the word 'expressly' in the same light with the gentlemen who opposed him; he thought every power to be expressly given that could be clearly comprehended within any accurate definition of the general power."[182] Differing understandings of "expressly" aside, Tucker obviously thought that without the term in the Amendment, the Constitution would allow exercise of implied powers and not very clearly implied powers at that.

The records of the other two abortive attempts to insert "expressly," in the House on August 21 and in the Senate on September 7, are more sketchy, with only the notation that when Elbridge Gerry moved the change in the House, he regarded it "of great importance."[183]

The alteration in Madison's wording which met success—that is, the inclusion of "or to the people" at the end of the Amendment—likewise had primarily a declaratory meaning to those in Congress. The addition evidently had its origin in the first of the amendments Madison proposed on June 8. This, probably based on suggestions from Virginia, New York, and North Carolina,[184] was "[t]hat there be prefixed to the constitution a declaration that all power is originally vested in, and consequently derived from, the

181. *Id.*

182. *Id.* at 1118 (*Annals of Cong.,* August 18, 1789).

183. *Id.* at 1127 (*Annals of Cong.,* August 21, 1789), 1150-51 (*Senate Journal,* September 7, 1789).

184. See remarks of William Smith in *id.* at 1076 (*Annals of Cong.,* August 14, 1789). Although not ratifying the Constitution, North Carolina's first convention had proposed amendments to it. See 4 Elliot, Debates 242-47, 251 (August 1, 2, 1788).

people."[185] The words "We the People" in the existing preamble, combined with the fact of ratification by conventions drawn from the people, made the prefix nugatory in the eyes of some. Madison replied that although self-evident, the words would have a calming effect, and that one would "be puzzled to find a better place" for them.[186] However, once the House voted to arrange the amendments in a separate and subsequent section rather than incorporate them into the body of the Constitution as Madison had originally contemplated,[187] his first proposal was dropped.[188] Thomas Tucker had already moved unsuccessfully to preface the reserved powers amendment with the wording "all powers being derived from the people," holding that this was "a better place to make this assertion than the introductory clause of the constitution . . . ."[189] Now that Madison's amendment to the preamble had been discarded, a comparable declaration in the reserved powers amendment did not create redundancy in wording, and so Congress added "or to the people."[190]

Although primarily declaratory, the final phrase of the Amendment may have had the further purpose of clarifying and qualifying the reservation of power to the states. The conclusion here is speculative, but seems plausible despite scanty records. As already described, the ratification debates in the states disclosed many still accepted the notion of legislative or governmental sovereignty. These people could have easily equated a reservation to the "states" with a reservation to the "state governments." Inserting "or to the people" guarded against this meaning.

VI. The Original Understanding of the Tenth Amendment

The course of the Tenth Amendment prior to its submission to the states constituted an essential step in the causal chain which led

185. Doc. Hist. of Bill of Rts. 1026 (*Annals of Cong.*, June 8, 1789).
186. *Id.* at 1076-77 (*Annals of Cong.*, August 14, 1789).
187. See note 176 *supra.*
188. Doc. Hist. of Bill of Rts. 1126 (*Annals of Cong.*, August 19, 1789).
189. *Id.* at 1118 (*Annals of Cong.*, August 18, 1789).
190. See notes 177-78 *supra* and accompanying text.

to its inclusion in the Constitution. For this reason, if no other, its evolution to September 1789 needs attention in relating its origins. But the pre-ratification history also gives insight into the Amendment's likely meaning to the state legislators who ratified it; and this insight is invaluable in the absence of other evidence.

The Amendment as sent to the states contrasted sharply with the Confederation's Article II and all but one of the state recommendations. It omitted reference to state sovereignty, and it failed to restrict the general government's powers to those "expressly" or "clearly" delegated. Against the background of 1787-1788, these silences must have caught the attention of contemporaries. Arguments for the sovereignty of the people surely sensitized politically aware Americans to the Constitution's claim to a base independent of the existing state governments. Similarly, debates over its specific provisions–especially the general welfare, necessary and proper, and supremacy clauses–surely sensitized Americans to the potential for a central government under the Constitution with an expansive reach, whatever the locus of sovereignty. Probably many shared the reaction of Virginia's two Senators to the amendments Congress sent to the states: "It is impossible," they wrote, "for us not to see the necessary tendency to consolidated Empire in the natural operation of the Constitution, if no further Amended than as now proposed."[191]

In particular, omission of "expressly" must have loomed as ominous to those who still looked to the states for their political salvation as it appeared hopeful to those who doubted the states' capacity to provide stable yet energetic government. Madison, it will be recalled, had argued in Congress that its *inclusion* would require the Constitution "to recount every minutia." By this reading of the term, its *exclusion* meant that the central government's powers needed no detailed base in the Constitution. Thomas Tucker had, of course, disagreed with Madison, holding that the term's inclusion would only require specification of powers in a clear and unambiguous fashion, but not in full detail. By this less restrictive reading of the word, exclusion of "expressly" was yet

191. Letter from Richard Henry Lee and William Grayson to the Speaker of the Virginia House of Representatives, September 28, 1789, in 2 THE LETTERS OF RICHARD HENRY LEE 507, 508 (J. Ballagh ed. 1914).

more significant. For central powers to exist, not only need the Constitution not describe them in full detail; the document need not even state them with great clarity. Moreover, each definition had eighteenth century currency.[192]

The preamble Congress attached to the twelve amendments sent to the states noted the desire in a number of the state conventions of 1787-1788 "that further declaratory and restrictive clauses should be added."[193] Contemporaries who knew of Article II of the Confederation and who had witnessed the earlier state and congressional debates certainly did not find the Tenth Amendment restrictive; it was instead declaratory of the Constitution's overall scheme. Indeed, in historical context, it probably reaffirmed the centralizing tendencies of the new system.

Still, it does not follow that the Amendment provided a constitutional warrant for limitless central authority: the constitutional scheme it attested was not one of unrestricted power. However much the nationalists in the Philadelphia Convention may have desired a consolidated government, they and their cohorts in the states subsequently denied that the Constitution established one. On its face, in fact, the document recognized the states. George Sutherland was doubtlessly correct when he argued that had Americans in 1787-1788 believed the Constitution would "reduce them

192. See 3 OXFORD ENGLISH DICTIONARY 448. W. CROSSKEY, POLITICS AND THE CONSTITUTION IN THE HISTORY OF THE UNITED STATES 680-84 (1953) contains a valuable discussion of the exchange between Tucker and Madison and of the eighteenth century meaning of "expressly." Crosskey's general discussion of the Tenth Amendment (see *id.* at 675-708) is useful on the Amendment's legislative history but otherwise is subservient to the author's overall thesis that the Constitution grants Congress a general power of legislation. Among other things, in contrast to his discussion of "expressly," Crosskey assigns overly narrow meanings to key words of the Amendment when the occasion demands. A crucial example is his treatment of "reserved." Contrasting it to "retained," Crosskey reads it in the narrow legalistic sense of creating a new right in the course of conveying other rights (as in reserving an easement through property being conveyed to another). By this reading, the Tenth Amendment arguably makes all state authority dependent on the Constitution. See *id.* at 701-02. This was *one* definition of the term in the eighteenth century, but not the only one (see 8 OXFORD ENGLISH DICTIONARY 513-14), and the disputants in the ratification debates—even the Antifederalists, who of course sought to preserve the states—commonly used the term interchangeably with "retained." See, *e.g.*, THE FEDERALIST No. 45; *Pa. Ratification Debates* 629 (dissent of the Pennsylvania minority).

193. DOC. HIST. OF BILL OF RTS. 1164 (Senate Journal, September 26, 1789).

[the states] to little more than geographical subdivisions of the national domain, . . . it would never have been ratified."[194] Ultimately the supporters of the Constitution had to defend it as partly federal and partly national, whether or not they used that precise formulation; and if, as seems the case, the Tenth Amendment is declaratory rather than restrictive, then it is declaratory of that proposition.

VII. Conclusion: Some Present Challenges

How, then, does the "original" Tenth Amendment square with Justice Rehnquist's recent glosses upon our so-called "federal" system–that is, upon what Americans in the late eighteenth century described as a mixed federal and national system? Beyond that, are the origins of the Amendment of any "use" to the present?

Read together, Rehnquist's dissenting opinion in *Fry v. United States* and his majority opinion in *National League of Cities v. Usery* show him arguing that the commerce clause provides no constitutional base for congressional action in instances where the action "transgresses an affirmative limitation on the exercise of its power . . . ."[195] One such affirmative limitation for Rehnquist is the Tenth Amendment, which "expressly declares" the Constitution's commitment to state freedom from impermissible congressional regulations.[196] A less explicit limitation derives from the American constitutional structure–in other words, from the relationship of the states and central government within the system erected by the Constitution. In *Fry,* Rehnquist spoke of the "concept of constitutional federalism."[197] In *Usery,* he relied on two 1869 cases in which the Court had held that "the Constitu-

194. Carter v. Carter Coal Co., 298 U.S. 238, 296 (1936).

195. *Usery* at 841. See *id.* at 842; *Fry* at 552-54. In this final section I am drawing freely on Lofgren, *National League of Cities v. Usery: Dual Federalism Reborn,* 4 Claremont J. of Public Affairs 19 (1977).

196. *Usery* at 843 (quoting *Fry* at 547 n. 7). See *id.* at 842-43; *Fry* at 550. See also Maryland v. Wirtz, 392 U.S. 183, 205 (1968) (Douglas, J., dissenting).

197. *Fry* at 554. See also Maryland v. Wirtz, 392 U.S., 183, 204 (1968) (Douglas, J., dissenting) ("constitutional principles of federalism").

tion, in all its provisions, looks to an indestructible Union, composed of indestructible states,"[198] and that "in many Articles of the Constitution the necessary existence of the States, and, within their proper spheres, the independent authority of the States, is distinctly recognized."[199] The Court had "repeatedly recognized," he found, "that there are attributes of sovereignty attaching to every state government which may not be impaired by Congress, not because Congress may lack an affirmative grant of legislative authority to reach the matter, but because the Constitution prohibits it from exercising the authority in that manner." Congress cannot control those decisions of a state which are "functions essential to [the state's] separate and independent existence." Regarding such decisions, state authority is "plenary."[200]

The Tenth Amendment argument is easily disposed of—at least superficially. By the historical evidence, the Court was correct in 1941 in labeling the Amendment a "truism."[201] Declaratory of the overall constitutional scheme, it had no independent force as originally understood. Yet the overall constitutional scheme did entail essential roles for the states. Accordingly, although merely declaratory, the Amendment militates at a deeper level against the central government's compromising the states' constitutional roles. As a beginning in this regard, one could do worse than to recall the rule of constitutional interpretation suggested by the Antifederalist "Brutus" (who, however, feared its application would lead to consolidated government): ambiguities in the Constitution are to be resolved in light of the general constitutional design.[202]

Such a rule of interpretation clearly has present-day ramifications: the states remain pretty much essential to today's constitu-

198. Texas v. White, 7 Wall. 700, 725 (1869), quoted in *Usery* at 844.

199. Lane County v. Oregon, 7 Wall. 71, 76 (1869), quoted in *Usery* at 844.

200. *Usery* at 845-46, in part quoting Coyle v. Smith, 221 U.S. 559, 580 (1911), in turn quoting Lane County v. Oregon, 7 Wall. 71, 76 (1868). See also Trimble v. Gordon, 97 S. Ct. 1459, 1469 (1977) (Rehnquist, J., dissenting), in which Rehnquist offers a short summary of his view of the federal system.

201. United States v. Darby, 312 U.S. 100, 124 (1941), quoted in text accompanying note 11 *supra*. *See* generally Berns, *The Meaning of the Tenth Amendment* in A NATION OF STATES 139, *passim* and esp. 144-51 (R. Goldwin, ed. 2d ed. 1974).

202. Robert Yates, *Letters of Brutus,* No. XI, January 31, 1788, excerpted in MASON, *supra* note 51, at 107, 108-09.

tion of government, which is to say that American government and politics would be significantly different without them. As an English scholar has observed, the states may not be sovereign in an ultimate sense, but even in the face of growing central authority they still exercise actual powers which "are of a different order from those of English local government authorities. Indeed, many of the powers they exercise *are* ones which the Englishman does automatically connect with sovereign legislative bodies; they are essentially 'central' powers—powers even of life and death."[203] And not only do the states have significant powers in law which are *separate from* those of the central government; they also figure as states *in* the central government and the politics associated with it. The Senate, the electoral college (and, within it, the practice of unit voting), and the "national" party system all contain elements that Madison and his contemporaries would have termed "federal." Put differently, "federalism" in its eighteenth century sense—which is what Justice Rehnquist seemed intent on defending, albeit without retaining the more precise (and perhaps more analytically useful) eighteenth century terminology—persists as both a legal and a political force.[204] Given this legacy, the states deserve constitutional recognition in areas where the explicit constitutional warrant for central power is obscure.

All of this, however, does not redeem Rehnquist's *application* of his structural argument in *Usery*. Neither the Tenth Amendment nor the constitutional system produces that "invisible radiation" Justice Holmes asked about in *Missouri v. Holland*.[205] Once an activity is conceded to affect interstate commerce, it is hard to get around Professor Corwin's observation that "the supremacy clause [does not] recognize any distinction between laws of a

203. M. VILE, THE STRUCTURE OF AMERICAN FEDERALISM 4 (1961).

204. See Diamond, *The Federalist on Federalism: "Neither a National Nor a Federal Constitution, But a Composition of Both,"* 86 YALE L.J. 1273, esp. 1281-85 (1977); Wechsler, *The Political Safeguards of Federalism: The Role of the States in the Composition and Selection of the National Government,* 54 COLUM. L. REV. 543 (1954) (but note that some "political safeguards" have disappeared or diminished in effectiveness since Professor Wechsler published this essay and that his terminology is less precise than Madison's).

205. 252 U.S. 416, 433-34 (1920). See generally Lofgren, *Missouri v. Holland in Historical Perspective,* 1975 SUPREME COURT REVIEW 77.

State which must yield and those which need not yield to a conflicting exercise by Congress of its delegated powers."[206]

What a review of the origins of the Tenth Amendment does suggest is that reconciling central authority with state autonomy has roots as deep as the Constitution. Untangling these roots reveals a willingness on the part of at least some late eighteenth-century Americans to set aside preconceived theories in order to gain acceptable government and to develop corresponding new theoretical formulations. If new distinctions and limits need to be developed and observed today (rethinking the constitutional meaning of "commerce" comes readily to mind),[207] then the innovations in constitutional thought during the 1780s provide a standing challenge.

206. CORWIN, *supra* note 6, at 257. See *id.* at 255-57. But cf. Tribe, *Unraveling National League of Cities: The New Federalism and Affirmative Rights to Essential Governmental Services,* 90 HARV. L. REV. 1065 (1977) (arguing that whatever were Rehnquist's and the Court's "motives," *Usery* gives recognition to affirmative individual rights which the central government may not infringe when such rights are furthered by the activities of states).

207. See generally BARBER, *supra* note 5, at 171-73 (questioning congressional use of the commerce power for "pretextual" purposes).

4

THE TREATY POWER: *Missouri* v. *Holland in Historical Perspective*

In April 1920 the Supreme Court rejected state claims based on the Tenth Amendment and upheld federal legislation which implemented a treaty to protect migratory birds. *Missouri v. Holland*[1] has since become, in the words of Professor Henkin, "perhaps the most famous and most discussed case in the constitutional law of foreign affairs."[2] Foreshadowed by debate prior to the decision, the dispute over *Holland* grew especially heated in the 1950s and has never entirely subsided. Some critics have read the Holmes opinion as creating a license for unlimited federal authority through the treaty power; more favorable commentators have argued that the decision did no more than ratify existing law.[3] By

1. 252 U.S. 416 (1920).

2. HENKIN, FOREIGN AFFAIRS AND THE CONSTITUTION 144 (1972) [hereafter HENKIN].

3. For the early controversy, see Boyd, *The Expanding Treaty Power,* in A.A.L.S., SELECTED ESSAYS ON CONSTITUTIONAL LAW 410, 422-28 (1938); Note, 33 HARV. L. REV. 281 (1919); Note, 8 CALIF. L. REV. 177 (1920); Note, 20 COLUM. L. REV. 692 (1920); Note, 29 YALE L. J. 445 (1920) (all supporting the position taken by the Court in *Holland*); Thompson, *State Sovereignty and the Treaty-Making Power,* 11 CALIF. L. REV. 242, 247-51 (1923); Black, Missouri v. Holland*—A Judicial Milepost on the Road to Absolutism,* 25 ILL. L. REV. 911 (1931); Note, 6 CORN. L.Q. 91 (1920); Note, 68 U. PA. L. REV. 160 (1920) (all adversely critical). Professor Corwin noted *Holland's* significance without overt approval, in *Constitutional Law in 1919-1920, II,* 15 AM. POL. SCI. REV. 52, 52-54 (1920), but had earlier supplied his imprimatur to similar arguments

examining *Holland* in historical context and by assessing its impact on subsequent constitutional law, this article seeks to shed some additional light on its meaning and on its capacity for raising controversy.

I. THE DEVELOPMENT OF FEDERAL MIGRATORY BIRD LEGISLATION

The active effort in Congress to gain federal protection for migratory birds dates to a bill introduced in 1904. Each succeeding Congress saw one or more similar bills introduced, until, on 3 March 1913, a measure passed both houses as an amendment to the Department of Agriculture Appropriation Act and received President Taft's signature.[4] State and national conservation organizations, progressive sportsmen's associations, state game officials, and members of the federal Department of Agriculture had backed the legislation. These groups feared for the birds' survival in the face of large-scale shooting not only of game birds but of such insectivorous birds as the robin, particularly during the spring breeding season. State protective legislation existed, but "strong temptation pressing upon every State to secure its full share of edible game birds during the spring and fall migrations . . . rendered harmonious and effective State supervision impossible."[5] Moreover, testimony established that the endangered birds destroyed insects which otherwise would damage millions of dollars of crops each year.[6]

---

in NATIONAL SUPREMACY: TREATY POWER V. STATE POWER (1913). Professor T. R. Powell likewise withheld open expression of judgment but seemed to think that the decision opened the way to constitutional amendment via treaties. See Powell, *Constitutional Law in 1919-1920,* 19 MICH. L. REV. 1, 11-13 (1920).

For the debate of the 1950s, see notes 219-31 *infra* and accompanying text.

The present legal status of the *Holland* case is set out in HENKIN 137-48, 383-96.

4. See Palmer, *Memorandum Concerning the Movement in Favor of Federal Protection,* in PROTECTION OF MIGRATORY AND INSECTIVOROUS GAME BIRDS OF THE UNITED STATES, H.R. REP. No. 680, 62d Cong., 2d Sess. 4-5 (1912); 49 CONG. REC. 4799 (1914); 37 STAT. 828, 847.

5. S. REP. No. 675, 62d Cong., 2d Sess., 1 (1912).

6. See *id., passim;* H.R. REP. No. 680, note 4 *supra, passim;* H.R. REP. No.

Professing agreement with its objectives, opponents of federal legislation mainly disputed its constitutionality. Migratory bird protection not only exceeded the federal government's delegated powers, they contended; it also clearly fell within the police power reserved to the states.[7]

The opposition's certainty in these regards was not matched by the advocates of protection. Senator George McLean, the sponsor of the bill which passed in 1913, was so unsure at first whether the legislation could stand on its own that he introduced a constitutional amendment to validate it. Later reversing himself on the need for an amendment, he still could not cite a specific constitutional clause to which the proposed legislation could be tied. Instead, he theorized that the national government held "implied attributes of sovereignty" with respect to those objects which the states acting alone could not achieve.[8] But this was a weak support in view of the Supreme Court's decision in *Kansas v. Colorado,* where Justice Brewer, writing for the Court, had explained:[9]

> The powers affecting the internal affairs of the states not granted to the United States by the Constitution, nor prohibited by it to the States, are reserved to the States respectively, and *all powers of a national character which are not delegated to the National Government by the Constitution are reserved to the people of the United States.*

McLean's exposition was otherwise dubious. It rested partly on a strained interpretation of several cases which had held that the states, acting in their sovereign capacities, were trustees for their citizens of animals *ferae naturae* within their boundaries. In these,

---

1424, 62d Cong., 3d Sess. *passim* (1913); Pearson, *The Federal Government to Protect Migratory Birds,* 24 CRAFTSMAN 395 (1913); Gladden, *Federal Protection for Migratory Birds,* 62 OUTING 345 (1913); *Weeks-McLean Law,* 15 BIRD LORE 137 (1913).

7. See 48 CONG. REC. 8547-49 (Rep. Mondell, 1 July 1912); 49 *id.* at 2725-27, 4330, 4337-39 (Reps. Mondell, Bartlett, Cox, 7, 28 Feb. 1913). *Cf. id.* at 1493 (Sen. Reed, 14 Jan. 1913). See also citations in note 16 *infra* for similar constitutional attacks on appropriations for bird protection after the legislation passed.

8. See 47 CONG. REC. 2564 (28 June 1911); 49 *id.* at 1489-94 (14 Jan. 1913). *Cf. id.* at 4331-32 (Rep. Lamb, 28 Feb. 1913).

9. Kansas v. Colorado, 206 U.S. 46, 90 (1907). (Emphasis added.)

the Court had in effect reserved decision on the extent of state power in the face of positive federal action.[10] To make his point, McLean equated such reservations with specific holdings favorable to federal power.[11] He also relied on an article by Senator George Sutherland in which the future Justice strove to show that dual federalism in domestic affairs and plenary national power in foreign affairs could constitutionally coexist.[12] By selective quotation, McLean portrayed Sutherland as arguing for plenary national power in both arenas, whenever the states were severally incompetent.[13] All told, the frankness of another supporter of protective legislation was much in order: "I do not know whether [the bill] is constitutional," he allowed, "but I do know that it is eternally right and in the end right will prevail."[14]

Passage of the Migratory Bird Act on 3 March 1913 failed to end the constitutional dispute. Even as some enforcement and evidently much voluntary compliance had a salutary effect on the bird population,[15] constitutional doubts figured in attacks on appropriations for continued enforcement.[16] Significantly, the De-

10. See Manchester v. Massachusetts, 139 U.S. 240, 264-66 (1891); Geer v. Connecticut, 161 U.S. 519, 528 (1896); The Abby Dodge, 223 U.S. 166, 173-75 (1912) (all mentioned by Sen. McLean, 49 CONG. REC. 1490).

11. See 49 CONG. REC. 1490 (14 Jan. 1913). Indeed, the opposition read the case law quite differently. See *id.* at 4337 (Rep. Sisson, 28 Feb. 1913).

12. See INTERNAL AND EXTERNAL POWERS OF THE NATIONAL GOVERNMENT, S. Doc. No. 417, 61st Cong., 2d Sess. *passim* (1910), also printed in 191 NORTH AM. REV. 373 (1910).

13. See 49 CONG. REC. 1491.

14. *Id.* at 4332 (Rep. Weeks, 28 Feb. 1913). The favorable House and Senate Reports on the bird protection bill further illustrate the uncertain grounds on which the legislation rested in the eyes of its proponents. While the House Report pictured the "interstate bird" as being in interstate commerce, it also included an admission from a supporter of bird protection that doubts about the bill's constitutionality plagued some who were "strongly in favor of [its] purpose." H.R. REP. No. 680, note 4 *supra,* at 2, 4-5. The Senate Report found support in the allegedly (but dubiously) analogous "power of the Federal Government to regulate by treaty the taking of migratory seals and fish," which of course overlooked the fact that no bird treaty yet existed. S. REP. No. 675, note 5 *supra,* at 2.

15. See the ANNUAL REPORTS OF THE DEPARTMENT OF AGRICULTURE: For 1915, at 246 (1916); for 1916, at 251-52 (1917); for 1917, at 265 (1918); for 1918, at 273-74 (1919).

16. See 51 CONG. REC. 8349-58, 8423-54 (Senate, 9, 12 May 1914); *53 id.* at 6763-70, 6804-06 (House, 24, 25 April 1916): *id.* at 10682-98 (Senate, 10 July

partment of Agriculture, which held responsibility for enforcement, showed little eagerness for a court test of the legislation.[17] And when prosecutions did occur, two federal courts handed down decisions adverse to the Act,[18] with these in turn giving new ammunition to its congressional opponents.[19]

Faced by these doubts and attacks, the birds' defenders called for a bird protection treaty with other North American nations. Indeed, the day the Senate passed the 1913 protection bill, Senator McLean had guardedly questioned whether "Congress can in the absence of a treaty exercise control over migratory birds or fishes."[20] Senator Elihu Root proceeded to introduce a resolution recommending negotiation of a treaty.[21] He remarked that "it may be that under the treaty-making power a situation can be created

---

1916); 54 *id.* at 968-70 (House, 6 Jan. 1917). Not all opponents of the law, however, opposed appropriations to enforce it. One argued: "If this law had been rigidly enforced, there would have been not three judicial decisions against its constitutionality, but a score. . . . Of course, the Supreme Court will overturn this law when the court reaches it; and the surest way to have the law pronounced unconstitutional, the most certain and effective way to arouse public sentiment against the effort to extend Federal police power throughout the country, is to enforce the law. Therefore I am in favor of the appropriation." 53 *id.* at 6804-05 (Rep. Mondell, 25 April 1916). For background, from a Progressive perspective, on the drive against bird protection after the 1913 law was enacted, see Phillips, *The Missouri Campaign: How the Middle West Has Organized to Defeat the Federal Migratory Bird Law,* 69 OUTING 77 (1916).

17. "The prosecution of cases arising under the law is under the jurisdiction of the Department of Justice. So far no case has been presented to this department [*i.e.,* Agriculture] which our solicitor has deemed it advisable to present to the Department of Justice." Letter from Secretary of Agriculture David F. Houston to Senator Robinson, 24 April 1914. 51 CONG. REC. 8350-51 (1914). "As to the matter of testing the law, I personally have no desire to press the matter. The only question is whether it can be kept out of the courts. There is pressure on the Department of Justice to have the law tested." Letter from Houston to Senator McLean, 23 April 1914. *Id.* at 8355. See generally *id.* at 8350-55, 8450-53 (Senate debates of 9, 12 May 1914, discussing the government's reticence over a court test of the 1913 law).

18. United States v. Shauver, 214 F. 154 (E.D. Ark. 1914); United States v. McCullagh, 221 F. 288 (D. Kans. 1915).

19. See 53 CONG. REC. 6763-69 (House, 24 April 1916); *id.* at 10682-93 (Sen. Reed, 10 July 1916); 54 *id.* at 968 (Rep. Doolittle, 6 Jan. 1917); 55 *id.* at 5547-48 (Sen. Reed, 30 July 1917); 56 *id.* at 7363, 7365 (Reps. Huddleston, Graham, 4 June 1918).

20. 49 *id.* at 1490 (14 Jan. 1913).

21. *Id.* at 1494.

in which the Government of the United States will have constitutional authority to deal with this subject."[22] Nothing came of Root's resolution, but in the next session one by McLean passed the Senate.[23]

In the treaty-related discussions during the appropriation debates, and in the debates on implementing the treaty which was quickly concluded with Great Britain (acting on behalf of Canada) and approved in 1916,[24] two general constitutional positions emerged. One held that bird protection, being a legitimate object of international concern, was a proper subject for a treaty, and that legislation implementing such a treaty was therefore constitutional because the treaty had status as supreme law and because the legislation violated no constitutional prohibition.[25] Opponents denied that a treaty could validate otherwise unconstitutional legislation, particularly when the legislation trenched on state police powers protected by the Tenth Amendment.[26] But after approval

22. Quoted by Sen. Robinson in 51 *id.* at 8349 (9 May 1914). This remark was omitted from the permanent edition of volume 49 of the *Congressional Record,* but from the context of Robinson's use of it, there seems no reason to doubt its accuracy.

Proponents of protection did not see a treaty merely as a means to remedy possible constitutional defects in the 1913 legislation. They also believed that international effort was needed for its real protective benefits. Finley, *Uncle Sam, Guardian of the Game,* 107 OUTLOOK 481, 487 (1914).

23. 50 CONG. REC. 2339-40 (7 July 1913). Forty years later, John W. Davis, who had been solicitor general from 1913 to 1918 under President Wilson, took credit for having suggested, in a conversation with Secretary of State Robert Lansing, the idea of concluding a treaty in order to establish a constitutionally viable basis for bird protection. See *Report of the Committee on International Law,* N.Y. State Bar Assoc., 30 Jan. 1953, at 60, quoted in *Hearings on S.J. Res. 1 before a Subcommittee of the Senate Committee on the Judiciary,* 84th Cong., 1st Sess. 139 (1955). Clearly, however, the idea antedated not only Lansing's tenure as secretary of state, but also Wilson's presidency. See notes 20-22 *supra* and accompanying text.

24. Convention with Great Britain for the Protection of Migratory Birds, 39 STAT. 1702 (1916).

25. See 51 CONG. REC. 8352-53, 8447-49, 8452-54 (Sen. McLean, 9, 12 May 1914); 53 *id.* at 10698-99 (Sen. McLean, 10 July 1916); 56 *id.* at 7361-62, 7367-71, 7377 (Reps. Stedman, Temple, Miller, Small, 4 June 1918); H.R. REP. No. 243, 65th Cong., 2d Sess. 1-2 (1918). *Cf.* 54 CONG. REC. 970 (Reps. Platt, Raker, 6 Jan. 1917); 55 *id.* at 4400 (Sens. Lodge, Hitchcock, McLean, 28 June 1917).

26. See 51 *id.* at 8351-54, 8447-49, 8452-54 (Sens. Robinson, Borah, Reed, Gore, 9, 12 May 1914); 53 *id.* at 10698 (Sen. Reed, 10 July 1916); 54 *id.* at

and ratification of the 1916 treaty, the opponents were less certain of their grounds than they had been in attacking legislation unaided by a treaty.[27] This is not surprising, for the advocates of protection on balance now constructed the better argument, having the relevant case law and treatises more clearly on their side.[28]

Legislation implementing the Migratory Bird Treaty of 1916 passed the Senate by voice vote on 30 July 1917 and cleared the House almost a year later, on 6 June 1918, on a 236 to 49 roll call.[29] After minor differences in the two versions were resolved, President Wilson signed the measure into law on 3 July 1918. The act forbade the hunting, killing, or subsequent sale and shipment of the bird species covered by the treaty, except as allowed by regulations to be established by the Secretary of Agriculture.[30] These regulations were quickly issued on 31 July.[31]

II. The Migratory Bird Act of 1913 in the Courts

By the time the 1918 law took effect, prosecution under the earlier legislation had virtually halted because of doubts concerning

---

969 (Rep. Sisson, 6 Jan. 1917); 56 *id.* at 7363-66, 7450-51 (Reps. Tillman, Huddleston, Graham, Mondell, 4, 6 June 1918).

27. See, *e.g.*, 55 id. at 4815 (Sen. Reed, 30 July 1917, admitting that Congress might have some power to legislate "touching the matters named in the treaty"); 54 Cong. Rec. App. 308 (1917). Similarly revealing the opponents' uncertainty was the amendment to the Migratory Bird Treaty Bill unsuccessfully proposed by Rep. Bland: "[T]his act shall not become effective until the provisions of the treaty between the United States and Great Britain covering the question of migratory birds shall have been ratified or approved by the legislatures of all the States of the Union." 56 *id.* at 7460 (6 June 1918).

28. *Compare, e.g.,* 56 *id.* at 7361-62, 7367-71 (Reps. Stedman, Temple, Miller, 4 June 1918, favoring the bill) *with id.* at 7365-66, 7450-52 (Reps. Graham, Mondell, Huddleston, 4, 6 June 1918, opposing the bill). See also note 27 *supra.* Representative Miller in his comments defending the bill somewhat anticipated the organic view of constitutional development that Holmes was to advance in his opinion in *Holland. Compare* 56 Cong. Rec. 7371 *with Holland* at 433; see also note 134 *infra* and accompanying text.

29. See 55 Cong. Rec. 5548 (30 July 1917); 56 *id.* at 7461-62, 8430, 8462 (6, 28, 29 June 1918).

30. 40 Stat. 755 (1918).

31. *Report of the Solicitor, October 3, 1919,* in Annual Reports of the Department of Agriculture for 1919 490 (1920).

its constitutionality.[32] Although in April 1914 the 1913 act was upheld against a constitutional challenge in one trial court,[33] the next month in another district Judge Trieber ruled against the legislation on a demurrer to an indictment in *United States v. Shauver,*[34] and other adverse rulings followed.[35] Trieber's opinion attracted attention in the continuing constitutional debates in Congress and typified the other opinions.

Approaching the constitutional issue with caution, Trieber stated that a holding of unconstitutionality could come "[o]nly if the question is practically free from real doubt."[36] He also recognized "that the United States . . . possess[ed] what is analogous to the police power, which every sovereign nation possesses, as to its own property."[37] This concession took on importance because the Government initially conceded that the law found no support in the Commerce Clause[38] and instead rested its case on the grant in Art. IV, § 3, that "[t]he Congress shall have power to . . . make all needful rules and regulations respecting the territory or other property belonging to the United States."[39] Trieber refused, however, to accept the claim that even though the Constitution made no affirmative grant, the national government possessed "implied attributes of sovereignty" which allowed it to act "where the state is clearly incompetent to save itself."[40] He bolstered his assertion

32. In its five years of operation by 30 June 1918, game wardens of the Department of Agriculture had reported 1,132 violations of the 1913 act, but prosecution in all but 29 of these had been withheld pending the Supreme Court's disposition of United States v. Shauver, 214 F.2d 154 (E.D. Ark. 1914), which is described in the text *infra,* at notes 52-53. See Annual Reports of the Department of Agriculture for 1918 274 (1919). The growth of this backlog can be traced in the Annual Reports for 1915-1917. See also note 17 *supra.*

33. United States v. Shaw (D.S.D., 18 April 1914), unreported but mentioned in *The Federal Migratory-Bird Law in the Courts,* 16 Bird Lore 322 (1914), and alluded to in State v. McCullagh, 96 Kansas 786, 790 (1915).

34. 214 Fed. 154 (E.D. Ark. 1914).

35. United States v. McCullagh, 221 F. 288 (D. Kan. 1915); State v. Sawyer, 113 Maine 458 (1915); State v. McCullagh, 96 Kansas 786 (1915). The state court rulings came when defendants raised the federal act as a bar to state prosecution for violation of state game regulations.

36. 214 Fed. at 156.

37. *Ibid.*

38. *Id.* at 160.

39. *Id.* at 156.

40. *Id.* at 156-57.

of vintage dual federalism with a long excerpt from Justice Brewer's opinion in *Kansas v. Colorado*,[41] which, among other things, remarked that "this is a government of enumerated powers," and that "[t]his natural construction of the original body of the Constitution is made absolutely certain by the tenth amendment."[42]

Trieber's reaffirmation of the Tenth Amendment made constitutionally irrelevant a mere showing that the states severally could not protect the birds.[43] *Shauver* thus turned on whether migratory birds were property of the United States or of the individual states. The long-standing American rule, which had evolved from English law and had been elaborated in the leading case of *Geer v. Connecticut*,[44] was that animals *ferae naturae* were the common property of the people of the individual states, with the states acting as trustees over that property for their citizens. Even after people reduced game to their own possession (as by killing a bird), the resulting personal ownership was qualified by a continuing state interest in its subsequent shipment or disposal.[45] Trieber therefore concluded that Art. IV, § 3, provided no support for the legislation.[46]

Nevertheless, after Trieber observed that government counsel had not argued "that the power to enact such legislation exists under the commerce clause,"[47] the Government advanced a Com-

41. 206 U.S. 46 (1907).

42. Quoted in 214 Fed. at 157.

43. Trieber repeated this point, explaining that the expediency of the legislation did not remove the constitutional problems besetting it, because "[i]t is the people alone who can amend the Constitution to grant Congress the power to enact such legislation as they deem necessary." *Id.* at 160. The courts thus had to take the Constitution as they found it. *Ibid.* He distinguished the examples of federal regulation of such objects as interstate lotteries, food and drugs, mailable packages, and prostitution. In those and similar instances, federal involvement had been "upheld under some provision of the Constitution, either that of the Post Office Department, the commerce clause, the taxing power, or some other grant." *Id.* at 159.

44. 161 U.S. 519 (1896).

45. See *id.* at 522-34. Justice White's analysis in *Greer* is summarized in *Shauver*, 214 Fed. 157-59, with attention there also to the subsequent conforming cases of New York *ex rel.* Silz v. Hesterberg, 211 U.S. 311 (1908), and The Abby Dodge, 223 U.S. 166 (1912).

46. 214 Fed. at 156-59.

47. *Id.* at 156. Indeed, according to Judge Trieber's later opinion on the Government's motion for rehearing, rather than simply not advancing a Commerce

merce Clause claim in its motion for rehearing. If it was true, as Trieber had averred in his original opinion, that a bird while in one state was owned by that state's citizens in their collective capacity, and after passing to another state was likewise owned by the latter's citizens, then the bird was an article in commerce.[48] But this argument similarly failed to impress Trieber. Again quoting from *Geer,* in which a state game regulation had been unsuccessfully attacked as trenching on the federal commerce power, Trieber denied the motion for rehearing.[49]

In the absence of the Government's case file for *Shauver,* which cannot be located,[50] it is difficult to determine accurately how important Trieber's ruling was for the Justice and Agriculture Departments. Yet they must have attached some importance to the defeat: Assistant Attorney General E. Marvin Underwood came from Washington to argue the motion for rehearing.[51] When that failed, the Government appealed on writ of error to the Supreme Court, which heard arguments on 18 October 1915 and assigned the case for reargument during the October 1916 Term.[52]

It seems probable, however, that the Migratory Bird Treaty with Britain, along with the legislation expected to implement it, now entered the Government's calculations, for government counsel first moved to postpone reargument, and eventually to dismiss, which the Court did in January 1919.[53] By this time, the Justice

---

Clause claim, government counsel had "conceded that the act cannot be sustained under the commerce clause." *Id.* at 160.

48. See Brief for the United States on Motion for Rehearing, quoted at 214 Fed. 160-61. Besides advancing this new argument in its motion for rehearing, the Government reiterated its two claims that the law should be accorded a strong presumption of constitutionality and that the birds in question were property of the United States. Judge Trieber agreed with the first of these, just as he had in his first opinion, and he almost summarily found no reason to change his conclusion respecting the second. *Id.* at 160.

49. *Id.* at 161.

50. Letter from Assistant Attorney General Henry E. Peterson, by John L. Murphy, Chief, Government Regulations Section, Department of Justice, to the Author, 23 May 1974, in author's possession. The Department's case file for *Holland* is also missing. *Ibid.*

51. 214 Fed. at 160.

52. Annual Report of the Attorney General of the United States for 1916 42 (1916).

53. United States v. Shauver, 248 U.S. 594 (1919). The Supreme Court's file

and Agriculture Departments probably concluded they would be on firmer ground in arguing for the 1918 act. They may also have wished to avoid an outright defeat in the Supreme Court on the 1913 legislation. Such a defeat would not make the new law any easier to defend.

Support for this construction of the Government's strategy and reasoning may be derived from a review of its brief before the Supreme Court in *Shauver*;[54] this elaborated the Government's two key arguments in District Court, but still lacked persuasiveness. A review of the brief also explains further why the 1913 law failed in the lower courts, and it offers added insight into why the Government in *Holland* placed major emphasis on the sweep of the treaty power and why Justice Holmes in his opinion took the same route.

The first of the Government's arguments in *Shauver* again rested on Art. IV, § 3, giving federal authorities power to regulate United States property. Admitting control over animals *ferae naturae* rested with the sovereign in his capacity as trustee for the animals' collective owners, the Government nevertheless denied that the states severally stood as sovereigns *qua* trustees in relation to game not permanently within any one state. A state, it argued, could not effectively preserve property which resided within its borders for only a portion of the year.[55] Nor could the states act jointly to preserve game through interstate compacts without approval of Congress. Here the Government's brief had a curious twist, in light of the fact that the 1913 legislation lacked a treaty base:[56]

---

for *Shauver* (Supreme Court Appellate Case 24323, National Archives, Washington, D.C.) fails to disclose the government's reasons for requesting postponement of reargument. Letter from Mr. M. M. Johnson, Legislative, Judicial, and Fiscal Branch, Civil Archives Division, National Archives and Records Service, to the Author, 11 September 1974, in author's possession.

54. Not otherwise conveniently available, the Brief for the United States in *Shauver* [hereinafter cited as Shauver Brief] appears as Appendix A in Brief for Appellee, Missouri v. Holland, 252 U.S. 416 (1920), and also in 55 CONG. REC. 4816-18 (1917). Citations hereinafter for the Shauver Brief are to the *Holland* Brief's appendix.

55. See Shauver Brief at 48-52.

56. *Id.* at 52-53.

> By ratifying the Constitution, the people expressly delegated to the Federal Government the exclusive right to protect wild life by treaties with foreign nations, and at the same time withdrew from the States the right, without the consent of Congress, to make among themselves agreements for such purpose, thus vesting in Congress the ultimate control and protection of same. By thus stripping the States of all power to protect migratory wild life for the greater part of the time, and expressly granting such power to the Federal Government, the people, by necessary implication from these express grants and the nature of the property, changed their trustee and vested the title to all migratory animals *ferae naturae* in the Federal Government in trust for themselves, the people of the United States and the common owners of such animals.

The Government also stressed that ownership of game found its basis in common law, and that no such law respecting migratory animals had ever been declared in the United States, so far as it concerned the federal government's interest.[57] Existing cases either had involved animals resident solely in one state or, to the extent migratory animals were involved, had not raised the issue of their migratory character. In the absence of federal law, the Court had not reached the question of federal control in the face of the states' police powers.[58] The Court was now being asked in a case of first impression to declare in favor of federal ownership.

In short, the 1913 legislation did not seek to establish an independent federal police power, which had been disallowed in *Kansas v. Colorado,* or to trench on the states in the exercise of their police power over game, which had been recognized in *Geer.*[59] It "merely provide[d] 'needful rules and regulations' respecting *property* of the United States within the territory of the several States, a power daily exercised by the Federal Government."[60]

The Government's second line of argument stressed that denial of its position concerning federal ownership of migratory game—

57. *Id.* at 48.

58. *Id.* at 54-56, citing and quoting especially *The Abby Dodge, Geer,* and *Manchester.*

59. Shauver Brief at 57.

60. *Ibid.*

that is, affirmation of state ownership–led to the conclusion that such game was in interstate commerce, for when the game crossed state lines, ownership in property was then transferred.[61] Commerce, the Government admitted, was not susceptible to precise definition, but "the mere 'transit' of persons or property, independently of purchase, sale, or exchange, was such intercourse as falls within the meaning of the word 'commerce.' "[62]

Certain problems are apparent in the Government's position in its *Shauver* brief. True, existing cases had not reached the question of federal powers over migratory game, but overall they hardly supported the Government's position. In *Geer v. Connecticut,* for example, Justice White had detailed at length how the sovereign had come to be regarded as trustee for animals *ferae naturae* and how, in America, this sovereign authority had vested first in the colonies and then in the states.[63] It "remains in them [the states] at the present day," he concluded, "in so far as its exercise may not [be] incompatible with, or restrained by, the rights conveyed to the Federal government by the Constitution."[64] On its face, perhaps, that formulation left an opening for federal regulation, but in reality it merely returned the question to whether the Constitution affirmatively granted the federal government either express or implied power over game.

On this score, the Government's federal trusteeship theory was dubious, for it essentially repeated the constitutionally irrelevant claim that since the states severally could not protect game and hence could not act effectively as trustees, there must exist a federal power to do so. Resting on no evidence beyond simple assertion, the theory ignored Justice Brewer's remark in *Kansas v. Colorado* that where powers of a national nature had not been positively assigned to the federal government, they remained with the people.[65] Despite Justice White's passing reservation in *Geer*

61. *Id.* at 59-62.

62. *Id.* at 61.

63. 161 U.S. at 522-29. As already mentioned, *Geer* was the leading case; it summarized previous developments and was followed in subsequent cases. See *e.g.,* McCready v. Virginia, 94 U.S. 391 (1877); notes 45 *supra* and 67-70 *infra.*

64. 161 U.S. at 528.

65. 206 U.S. at 90. Although Hammer v. Dagenhart, 247 U.S. 251 (1918), had not yet been decided, it is similarly instructive to consider the Government's argu-

of the question of federal power, the Government theory also ran against White's actual emphasis in *Geer* on the states as trustees for animals *ferae naturae.*[66] And in the same year White had reiterated this emphasis when, speaking for the Court in *Ward v. Race Horse,*[67] he maintained "the complete power to regulate the killing of game within its borders" was "a necessary incident of . . . [state] authority." In *Patsone v. Pennsylvania,*[68] Justice Holmes had read *Geer* as establishing "the protection of wild life" as a "lawful [state] object" in the face of a claim that certain state regulations worked a deprivation of Fourteenth Amendment rights. Wild game, he said, was something "which the State may preserve for its own citizens if it pleases."[69] As if to ratify this line of interpretation, White, now the Chief Justice, observed later in the same Term in which *Shauver* was argued that "[i]t is not to be doubted that the power to preserve fish and game within its borders is inherent in the sovereignty of the state."[70]

Arguably, too, the Government's Commerce Clause argument

---

ment in *Shauver* in light of the Court's opinion in *Dagenhart.* See note 111 *infra* and accompanying text; note 139 *infra.*

66. 161 U.S. at 527-29.

67. 163 U.S. 504, 510 (1896). Later in the same opinion White reiterated: "The power of all the States to regulate the killing of game within their borders will not be gainsaid." *Id.* at 514.

68. 232 U.S. 138, 143 (1914).

69. *Id.* at 145-46. Like Justice White's argument in *Ward,* Holmes's narrow argument in *Patsone* was that if a treaty were to confer rights in derogation of a state's customary and well-established powers of police, it would have to do so explicitly and could not do so by implication.

70. New York *ex rel.* Kennedy v. Becker, 241 U.S. 556, 562 (1916) (citing *Geer* and *Race Horse*). Though delivered by Chief Justice White, this opinion had been written by Justice Charles Evans Hughes prior to his resignation. *Id.* at 559. The remainder of the statement quoted in the text was: "subject, of course, to any valid exercise of authority under the provisions of the Federal Constitution." *Id.* at 562. So here again the question of state power in the presence of valid federal regulation was reserved, but, like the Government brief in *Shauver,* see text at note 64 *supra,* at most this too only returned the issue to what constituted valid federal regulation. More specifically, in context, the Court doubtless meant to reserve the question with reference to the exercise of the federal government's treaty power, for the case involved (and the Court disallowed) an Indian claim that a federal treaty conveying hunting and fishing rights acted as a bar to state prosecution for violations of state game laws. Hence, although the Court had already heard arguments in *Shauver,* it probably was not referring to an exercise of federal authority lacking a treaty base, as in *Shauver.*

fell before *Geer*. There, as a bar to state prosecution, the defendant had claimed that the birds he had killed and sold out of state had entered interstate commerce. Not so, Justice White had held, finding that "the errors which this argument involves are manifest."[71] For one thing, personal ownership of such game remained qualified by the state's authority, under its trusteeship and police powers, to fix conditions on the killing and sale of game to such an extent that "it may well be doubted whether commerce is created" by those acts.[72] For another, even if it were conceded that the acts in question created commerce, "it [did] not follow that such internal commerce became necessarily the subject matter of interstate commerce, and therefore under the control of the Constitution of the United States. The distinction between internal and external commerce and interstate commerce is marked, and has always been recognized by this court."[73]

Of course, one could question whether White's remarks about commerce in *Geer* were relevant to the migratory bird legislation of 1913. The game involved in *Geer* had already been killed and reduced to personal possession within a particular state. The 1913 law, although regulating the killing of game within individual states, aimed to preserve live birds in their flights across state lines. Yet the Government could not cite a single case in which the item being regulated as interstate commerce had crossed state lines of its own volition and not as a result of its owner's will or act.[74] In seeking to establish "transit" as a sufficient condition to define "commerce," it could only "submit" that self-volition in movement was constitutionally irrelevant.[75]

71. 161 U.S. at 530.
72. *Ibid.*
73. *Id.* at 530-31.
74. See Shauver Brief at 61-62.
75. "Undoubtedly, under *Kelley v. Rhoades,* 188 U.S. 1, the Federal Government could prevent the owner from driving diseased cattle from the quarantined State into the other. It is submitted that the fact that the same cattle cross the quarantine line of their own volition instead of that of their owner does not so change their status as participants in interstate intercourse as to destroy the power of Congress to prevent their crossing the State line." *Id.* at 62. Later, to be sure, the Court held that freely ranging cattle which crossed state lines were in interstate commerce, but this conclusion rested on the argument that movement of the cattle was "made possible by the failure of their owners to restrict their

The remaining constitutional support the Government offered in *Shauver* was the Treaty Clause. The clear implication of the Government's brief was that the existence of the treaty power created an authority to regulate migratory birds independent of the actual conclusion of a treaty on the subject.[76] Yet, however prescient reference to the treaty power was with respect to the 1918 legislation, it provided no base for the 1913 legislation or for any action in the absence of a treaty. The necessity to resort to such sophistry simply underscores the constitutional infirmities which beset the Government's case within the doctrinal climate of the decade.

This critique of the Government's brief is not intended to suggest that no way existed by 1920 to uphold a bird protection law that lacked treaty support.[77] Under the guise of approving commercial regulation and taxation, the Court had already legitimated what in fact constituted federal police power in other areas.[78] The point instead is that, compared with the ease offered by the treaty route, the task clearly would have been a more difficult one, particularly in view of the willingness in 1918 of a five-man majority to strike down federal child labor legislation which rested on the Commercial Clause.[79]

---

ranging, and [was] due, therefore, to the will of their owners." Thornton v. United States, 271 U.S. 414, 425 (1926).

76. Shauver Brief at 52-53.

77. In 1916, for instance, Professor Corwin entered a vigorous defense of the 1913 act on grounds similar to those advanced by the Government in *Shauver*. Corwin, *Game Protection and the Constitution,* 14 MICH. L. REV. 613 (1916).

78. See, *e.g.*, Champion v. Ames, 188 U.S. 321 (1903) (lotteries); McCray v. United States, 195 U.S. 27 (1904) (colored oleomargarine); Hipolite Egg Co. v. United States, 220 U.S. 45 (1911) (pure food and drugs); Hoke v. United States, 227 U.S. 308 (1913) (prostitution).

79. Hammer v. Dagenhart, 247 U.S. 251 (1918); Henkin, *The Treaty Makers and the Law Makers: The Law of the Land and Foreign Relations,* 107 U. PA. L. REV. 903, 915 n.24 (1959). Justice Day's opinion for the Court in *Dagenhart* abounds with passages which could as easily have applied to bird regulation devoid of a treaty base. Consider, for example: "[T]he act in a twofold sense is repugnant to the Constitution. It not only transcends the authority delegated to Congress over commerce but also exerts a power as to a purely local matter to which the federal authority does not extend." 247 U.S. at 276. That the situation regarding federal regulation was becoming clouded is further indicated by United States v. Doremus, 249 U.S. 86 (1919), in which the Court, per Justice Day, upheld federal narcotics legislation by only a 5-to-4 vote.

III. The Migratory Bird Treaty Act of 1918: Judicial Preliminaries

The Migratory Bird Treaty Act of 1918, unlike its predecessor, immediately received favorable treatment in the federal district courts. In fact, the first decision, in *United States v. Thompson,*[80] came from Judge Trieber. The Government, evidently attaching considerable importance to the case, was represented in *Thompson* not only by the resident United States attorney but also by an assistant attorney general and by the solicitor of the Department of Agriculture.[81] Trieber reached his finding of constitutionality without noticeably more difficulty than had attended his opposite conclusion in *Shauver*. He at first stressed the wording of the Supremacy Clause: "when referring to treaties, the only limitation is 'which shall be made under the authority of the United States,' omitting the words 'in pursuance of the Constitution.' "[82] Later, though, he recognized other limitations on treaty-making.[83] While a court might therefore hold a treaty void, it should do so only " 'in a very clear case indeed,' "[84] and the treaty with Britain was not such a case, even though it involved a subject otherwise governed by state regulation.[85] The major contrast with *Shauver* was that the Tenth Amendment now did not apply. The Constitution specifically granted the treaty power to the federal government and forbade its exercise to the states.[86]

Although supported by opinions differing in emphasis and detail (and shorter as well), subsequent district court decisions were essentially similar.[87] That these later cases were prosecuted by

80. 258 Fed. 257 (E.D. Ark. 1919).

81. *Ibid.*

82. *Id.* at 258, and see 258-61.

83. *Id.* at 262, relying especially on Geofroy v. Riggs, 133 U.S. 258, 266-67 (1890).

84. 258 Fed. at 268, quoting Chase, J., in Ware v. Hylton, 3 Dall. 199, 237 (1796) (separate opinion).

85. *Compare* 258 Fed. at 268 *with id.* at 258-59.

86. *Id.* at 262-67.

87. United States v. Samples, 258 Fed. 479 (W.D. Mo. 1919); United States v. Selkirk, 258 Fed. 775 (S.D. Tex. 1919); United States v. Rockefeller, 260 Fed. 346 (D. Mont. 1919).

local United States attorneys without assistance on argument from Washington may indicate that the Government itself was confident of their outcome. Relatedly, the Bureau Chief in Agriculture charged with implementing the act claimed that the lower court victories "[had] removed to a large extent the doubt existing in some quarters concerning the validity of the act, and . . . [were] a decided deterrent to those inclined to violate the law."[88] Certainly the statistics on prosecutions and convictions, when compared with the virtual hiatus in enforcement under the 1913 act after the *Shauver* case had come before the Supreme Court, reveal a more active and confident enforcement policy.[89]

The case which eventually confronted the Supreme Court arose in Missouri when two local citizens were indicted for violating the new federal game regulations. This led the state to sue to enjoin further enforcement of the act by the federal game warden, Ray P. Holland. After argument on defendants' demurrers to their indictments and on the Government's motion to dismiss Missouri's bill in equity for an injunction against Holland, Judge Van Valkenburgh overruled the demurrers and dismissed the bill, but not without remarking that the 1918 Act would have been unconstitutional without its treaty base.[90] Missouri then appealed to the

88. ANNUAL REPORTS OF THE DEPARTMENT OF AGRICULTURE FOR 1919 at 295 (1920).

89. In the first year of the new law's operation, the Government obtained 110 convictions, and 393 in its second year. For these and similarly revealing figures on enforcement of the 1918 law, see the ANNUAL REPORTS OF THE DEPARTMENT OF AGRICULTURE; For 1918, at 274 (1919); for 1919, at 490-91 (1920); for 1920, at 599-600 (1921). There are minor discrepancies between the Department of Agriculture's statistics and those found in the *Reports of the Attorney General* for these years. I have used Agriculture's figures because they are more detailed. For figures on the 1913 law, see note 32 *supra*. Interestingly, about the time prosecutions under the 1918 law were beginning and the law was being upheld in the lower courts, the Supreme Court in May 1919 sidestepped a final opportunity to rule on the earlier 1913 act. When a South Dakota defendant raised the 1913 law as a bar to state prosecution for violation of state game laws, the State argued, *inter alia,* that the federal law was unconstitutional. The Court, per Justice Brandeis, upheld the defendant's state conviction by finding no conflict between the state and federal acts. Carey v. South Dakota, 250 U.S. 118 (1919).

90. 258 Fed. 479 (W.D. Mo. 1919). Although conceding that the Migratory Bird Treaty Act of 1918 would have been unconstitutional in the absence of the 1916 treaty, *id.* at 481, and recognizing that treaties inconsistent with the Constitution were invalid, *id.* at 482-83, the judge found that the treaty violated no con-

Supreme Court and was joined in its attack on the 1918 law by the State of Kansas as amicus curiae. The Association for the Protection of the Adirondacks filed an amicus brief supporting the legislation.[91]

The Missouri and Kansas briefs anticipated a major criticism of the eventual decision. If an enactment which is otherwise unconstitutional becomes constitutional when passed pursuant to a treaty, the states argued, then for all practical purposes the Constitution can be amended without resort to the formal amendment process.[92] Accordingly, to uphold the Migratory Bird Treaty Act would set the stage for the President and Senate, acting with the assistance of some obliging foreign power (at one point Missouri mentioned Turkey),[93] to invade all internal state affairs, regulate child labor, cede state territory, alter modes of elections, and even force the introduction of opium.[94] Missouri warned that "[w]hen the power of the states over their purely internal affairs is destroyed, the system of government devised by the Constitution is destroyed."[95]

Maintaining that treaties must be subordinate to the Constitution,[96] both states rejected the notion that the Supremacy Clause freed treaties from the need to conform to the Constitution. The clause admittedly spoke not of treaties made pursuant to the Constitution, but of "[t]reaties made, or which shall be made, under the Authority of the United States." The two states explained, however, that this phrasing derived only from the necessity in 1787 to validate preexisting treaties.[97] Included in the resulting limitations on treaties were the reservations incorporated in the Tenth Amendment.[98] One of the powers thereby reserved to the

---

stitutional provisions and involved a matter of mutual interest between nations. 258 Fed. at 483-85.

91. 252 U.S. at 430.

92. Brief for Appellant, esp. at 41-42, 63-73; Brief for Kansas at 29-37; *cf. id.* at 25-29, 37-71.

93. Brief for Appellant at 63.

94. Brief for Appellant at 63, 73; Brief for Kansas at 28-29.

95. Brief for Appellant at 64.

96. *Id.* at 43-59, 78-86; Brief for Kansas at 37-77.

97. Brief for Appellant at 83-84; Brief for Kansas at 31-32.

98. Brief for Appellant at 56-59; Brief for Kansas at 13, 40-45.

states, in both their trusteeship and police capacities, was the regulation of game.[99]

Like the opinions in the favorable decisions below, the Government in its brief on behalf of Holland found the Tenth Amendment "wholly irrelevant" where treaties were involved.[100] The power to make treaties had been expressly granted to federal authorities by the Constitution and hence was not reserved.[101] It agreed that treaties were subject to prohibitions found in the Constitution, but failed to enumerate those prohibitions and merely contended that no prohibition barred a treaty dealing with a subject of mutual interest and proper negotiation between nations.[102] Although conceding dual federalism characterized domestic federal-state relationships, the brief explained that "when we come to deal with national questions affecting the interests not only of our own country but of other countries as well, we confront a different situation."[103] This position, the Government noted,[104] was consistent with the Court's earlier ruling in *Geer,* which legitimated state regulation only "in so far as its exercise may not be incompatible with, or restrained by, the rights conveyed to the Federal Government by the Constitution."[105]

The issue thus became whether bird protection was a proper subject of international negotiation. It would "not admit of doubt," argued the Government, that protection was "a matter of very great importance to both countries."[106] In fact, because the birds crossed international borders, such protection was only possible through a treaty.[107] To illustrate the importance of the birds, the

99. Brief for Appellant at 30-44, 64-73; Brief for Kansas at 9-13, 57-59.

100. Brief for Appellee at 18.

101. *Id.* at 8-14, 24-27.

102. *Id.* at 33-34, 36-41.

103. *Id.* at 13-14.

104. *Id.* at 30-31.

105. 161 U.S. at 528, quoted *id.* at 31. Whether the Court in *Geer* specifically meant the exercise by the federal government of its treaty power is problematic, but in at least one more recent case, when it entered a similar reservation, the Court probably did specifically have in mind possible exercise of the treaty power. New York *ex rel.* Kennedy v. Becker, 241 U.S. 556, 562 (1916). See notes 69 and 70 *supra* and accompanying text.

106. Brief for Appellee at 42.

107. *Id.* at 42-43.

brief included data on their economic value in preserving crops and forests from insects.[108]

The Government also contended that the legislation of 1918 was valid even in the absence of a treaty.[109] However, it elaborated this point merely by including its earlier *Shauver* brief as an appendix,[110] which may suggest the Government suspected–and not without cause–that such an argument would likely have but marginal, if any, effect on the Court.[111] The Association for the Protection of the Adirondacks nevertheless gave the point greater emphasis, plus a new dimension. Perhaps recognizing that federal title to migratory birds was difficult, if not impossible, to establish, the Association massed a plethora of facts to demonstrate how birds were necessary to the preservation of national forests.[112] Federal bird regulation was therefore a necessary and proper means of regulating objects–that is, the forests–which were themselves undeniably federal property within the meaning of Art. IV, § 3.[113] Of course, this argument reinforced as well the Government's contention that the birds constituted a substantial national interest.

IV. Justice Holmes's Opinion for the Court: An Explication

The effort to put the federal government in the business of protecting migratory birds now squarely faced its final constitutional test. In outline, the Court's opinion, written by Justice Holmes, is simple enough. Quickly conceding Missouri's standing to sue,[114]

108. *Id.* at 63-79 (Appendix B).

109. *Id.* at 7-8.

110. *Id.* at 47-62 (Appendix A).

111. It should be recalled in this connection that the lower court in *Holland* had remarked that the 1918 law would have been unconstitutional without its treaty base. Perhaps also contributing to the Government's deemphasis of the non-treaty arguments was the Court's recent surprising and highly unexpected invalidation of the 1916 Child Labor Law in Hammer v. Dagenhart. On the surprised reactions to *Dagenhart,* see Wood, Constitutional Politics in the Progressive Era: Child Labor and the Law 169-70 (1968).

112. *See* Brief for Association for the Protection of the Adirondacks *passim.*

113. *Id.* at 11-28.

114. 252 U.S. at 431.

Holmes narrowed the issue to "whether the treaty and statute are void as an interference with the rights reserved to the States."[115] He next explored possible tests for determining the constitutional validity of treaties and pursuant statutes, concluding that they must involve matters of national interest and must not contravene specific constitutional prohibitions.[116] Finally, he applied these tests to the challenged treaty and statute and held them constitutional.[117] Only Justices Van Devanter and Pitney dissented, and without opinion.[118] But to argue that the majority opinion is a model of clarity would be to fly in the face of a half-century of dispute over its meaning. A close reading reveals the complexities of Holmes's handiwork.

Holmes focused his analysis on the treaty and not on the statute implementing it, for "[i]f the treaty is valid there can be no dispute about the validity of the statute under Article I, § 8, as a necessary and proper means to execute the powers of the Government."[119] Such a focus gave a nearly irrebuttable presumption of constitutionality to the challenged legislation. To be sure, the Tenth Amendment reserved to the states or the people those powers not delegated to the United States, but the Constitution expressly delegated the power to make treaties and declared that "treaties made under the authority of the United States, along with the Constitution and laws of the United States, made in pursuance thereof," were the supreme law of the land.[120] With such a foundation, the decision no longer turned on whether the substantive power to be exercised by a treaty and pursuant statute had itself been delegated. "The language of the Constitution as to the supremacy of treaties being general," Holmes declared, "the question before us is narrowed to an inquiry into the ground upon which the present supposed exception is placed."[121]

He thereupon explored the possible limits to a valid treaty. "One such limit," he wrote, "is that what an act of Congress could

115. *Id.* at 432.
116. *Id.* at 432-35.
117. *Id.* at 434-35.
118. *Id.* at 435.
119. *Id.* at 432.
120. *Ibid.*
121. *Ibid.*

not do unaided, in derogation of the powers reserved to the States, a treaty cannot do."[122] But whether or not the district courts had been correct in striking down the 1913 bird protection law as an invasion of the reserved powers of the states,[123] this limit ran afoul of the text of the Constitution by which "[a]cts of Congress are the supreme law of the land only when made in pursuance of the Constitution, while treaties are declared to be so when made under the authority of the United States."[124] That textual distinction led Holmes into the comment probably most responsible for the later criticism of *Missouri v. Holland:* "It is open to question whether the authority of the United States means more than the formal acts prescribed to make the convention."[125] By itself, this statement suggested that any treaty, when duly made, approved, and ratified, held status as supreme law, other constitutional provisions notwithstanding.

Holmes, however, had only raised a question. Although the question contained far-reaching implications, he promptly limited their reach by recognizing that limitations to the treaty power did exist.[126] Yet a distinction remained between treaties and legislation unaided by treaties, for the limitations to treaties "must be ascertained in a different way."[127] This, of course, was but a restatement of the view that while the federal government held no general power of legislation, it did hold a general treaty power. So Holmes to this point had rejected the argument that the alleged unconstitutionality of the Migratory Bird Act of 1913 doomed the 1918 act. He still faced the task of specifying the limits to the treaty power.

Perhaps taking his cue from the Government's argument, Holmes did not initially specify the outer limits of valid treaties, but instead offered, at least by implication, a threshold test which can be labeled the "national interest test." A constitutionally valid treaty,

122. *Ibid.*
123. See *id.* at 432-33. Holmes cited *Shauver* and *McCullagh.*
124. 252 U.S. at 433.
125. *Ibid.*
126. "We do not mean to imply that there are no qualifications to the treaty-making power." *Ibid.*
127. *Ibid.*

he implied, must deal with a national concern. His precise wording was this:[128]

> It is obvious that there may be matters of the sharpest exigency for the national well being that an act of Congress could not deal with but that a treaty followed by such an act could, and it is not lightly to be assumed that, in matters requiring national action, "a power which must belong to and somewhere reside in every civilized government" is not to be found.

But this formulation left open whether, in order to be appropriate objects for the treaty power, such matters of "sharpest exigency for the national well being"—that is, matters of national concern—need be susceptible only to actual international action for their solution. The passage can be read thus, or it can be read as defending the exercise of the treaty power for the primary purpose of dealing with matters of national concern susceptible to domestic solution but which are constitutionally beyond the unaided legislative power of Congress.[129]

His next sentence hinted that Holmes had in mind the second of these alternatives. Referring to *Andrews v. Andrews,*[130] which he had just cited and quoted, he wrote: "What was said in that case with regard to the powers of the States applies with equal force to the powers of the nation in cases where the States individually are incompetent to act."[131] In other words, because "a power which must belong to and somewhere reside in every civilized government" *had* to exist somewhere in American government, a treaty followed by a statute could serve to clothe the national govern-

128. *Ibid.*

129. For the view that the second alternative was Holmes's intended meaning, see, *e.g.*, Nicholson, *The Federal Spending Power,* 9 TEMPLE L.Q. 3, 5 n.10 (1934).

130. 188 U.S. 14 (1903). At issue there was whether a fraudulently obtained South Dakota divorce, which had not been challenged in South Dakota, must be recognized in Massachusetts pursuant to the Full Faith and Credit Clause of Art. IV, § 1. The Court, per Justice White, held that to require recognition would emasculate Massachusetts's control over an important area of internal police and thereby effectively destroy a power which must exist "in every civilized government." *Id.* at 32-33.

131. 252 U.S. at 433.

ment with needed plenary power that the Constitution otherwise failed to grant. That he thought a requirement for actual international action was not a requisite to valid exercise of the treaty power follows also from Holmes's later affirmation that even if state action would suffice, this would not withdraw the subject of bird protection from the treaty power.[132]

If such was Holmes's intended meaning, it was arguably dictum. He not only concluded that in reality state action was insufficient; he seems to have accepted the Government's argument that national action by itself was similarly insufficient and thus a truly international effort was in fact required.[133] Accordingly, he had no need to decide whether a matter not requiring international action, and not within the federal government's delegated powers, might still be confided in it by a treaty with some willing foreign power.

In any event, to lay the basis for his eventual conclusion that the Migratory Bird Treaty concerned a matter of national interest, Holmes had to establish that the category "national interest" was not a fixed, unchanging one. (One can imagine his anticipating the objection that the founding fathers would hardly have thought bird protection a significant enough problem to warrant a treaty.)[134] He argued:[135]

> [W]hen we are dealing with words that also are a constituent act, like the Constitution of the United States, we must realize that they have called into life a being the development of which could not have been foreseen completely by the most gifted of its begetters. It was enough for them to realize or to hope that they had created an organism; it has taken a century and has cost their suc-

132. *Id.* at 435. Because Holmes denied that state incompetency was necessary in order to commit a subject to the treaty power, I believe that Professor Powell was mistaken in writing, in his critique of *Holland,* that "[t]he fact that the states are individually incompetent to deal with the subject matter seems to be regarded as important [in establishing the existence of a national interest for treaty purposes]." Powell, note 3 *supra,* at 12.

133. 252 U.S. at 435.

134. During debate on the Migratory Bird Treaty Bill, one of its congressional supporters had anticipated such an argument and had proceeded to counter it with an argument similar in many respects to the one developed by Holmes at this point in *Holland,* 56 CONG. REC. 7371 (Rep. Miller, 4 June 1918).

135. 252 U.S. at 433.

> cessors much sweat and blood to prove that they created a nation. The case before us must be considered in the light of our whole experience and not merely in that of what was said a hundred years ago.

Besides conjuring up young Captain Holmes, thrice wounded in the Civil War, this language unambiguously displayed the Constitution as a flexible, dynamic instrument. It led, however, to a superficially more cryptic comment about the Tenth Amendment. "The only question," wrote Holmes, "is whether it [the treaty] is forbidden by some invisible radiation from the general terms of the Tenth Amendment."[136] Finally there came the sentence relating Holmes's comment about an organic constitution to Missouri's Tenth Amendment claim: "We must consider what this country has become in deciding what that Amendment has reserved."[137]

These last comments–about "invisible radiation" and the need to consider what the country has become–were phrased and positioned so that taking them out of context is easy.[138] This has further contributed to the controversy surrounding the opinion, because, out of context, the remarks surely conflict with the restrictive gloss Justice Brewer had put on the Tenth Amendment in *Kansas v. Colorado,* a gloss comparable to that put on it by the opponents of the earlier bird protection act of 1913, by Missouri in the instant case, and so recently by the Court in *Hammer v. Dagenhart.* Whether Holmes, a dissenter in *Dagenhart*[139] and a master of the

136. *Id.* at 433-34.

137. *Id.* at 434.

138. Although Holmes had sufficient warrant to refer to the Tenth Amendment's "invisible radiation," the phrase greatly disturbed those who saw the Amendment as a most concrete and highly visible bulwark of the federal system, and this probably contributed to their focusing narrowly on the phrase as derogatory of the Amendment itself. See, *e.g.,* Appellant's Petition for Rehearing at 9. Moreover, Holmes's remark that "[w]e must consider what this country has become in deciding what that amendment has reserved" was separated by most of a fairly long paragraph from the main body of his related comments about matters involving an expansible national interest, with which *treaties* could deal. 252 U.S. at 433-34.

139. 247 U.S. at 277. In conjunction with his remarks about the Tenth Amendment in *Holland,* consider, for example, this remark from Holmes's *Dagenhart* dissent: "I should have thought that the most conspicuous decisions of this Court had made it clear that the power to regulate commerce *and other constitutional powers* could not be cut down or qualified by the fact that it might interfere with

English language, deliberately phrased and positioned the remarks to achieve this appearance of conflict makes for intriguing speculation. Yet they must not be read as part of a general gloss on the Amendment. Indeed, had the 1920 Court so understood them, one suspects Holmes's opinion would have failed to command the 7 to 2 majority it did. They instead must be read more narrowly as part of an explication of the role of the Amendment in treaty adjudication.

For Holmes was still exploring the limits to valid treaties. Particularly, he was ruling on Missouri's Tenth Amendment claim. With this in mind, a sorting out of Holmes's thoughts reveals the following line of reasoning: Matters of national interest, and certainly those requiring international action for their solution, fall within the treaty power, a power which is expressly granted to the federal government, expressly denied to the states, and thus clearly excepted from the Tenth Amendment's restrictions. Because the category of national interest has grown with the nation, the permissible scope of treaties has likewise grown. This being the case, the growth of the nation has correspondingly and (to adapt Holmes's usage) visibly narrowed the reach of the Tenth Amendment, so far as treaties were involved. To be sure, a reversal of roles occurred in Holmes's argument. Contemporaries more commonly regarded the Tenth Amendment as the restricting, rather than the restricted, constitutional element. Nonetheless, since the Amendment did not visibly restrict treaties, all that remained was to inquire whether it had "some invisible radiation" which did so. Holmes gave no direct answer;[140] to ask the question was to answer it in the negative.

Holmes, however, did not stop at that point. Instead, in the midst of his remarks on the organic nature of the Constitution and on the inapplicability of the Tenth Amendment, he observed that the Bird Treaty "d[id] not contravene any prohibitory words to be found in the Constitution."[141] He thereby implied another and more general test for assessing the constitutional validity of trea-

---

the carrying out of the domestic policy of any State." 247 U.S. at 278. (Emphasis added.)

140. 252 U.S. at 434.

141. *Id.* at 433.

ties. This, which can be labeled the "no-conflict test," was perhaps required to square Holmes's organic conception of the Constitution with the more general American notion of constitutionalism in the sense of limited government. For if national interest were an expansible concept, then, by itself, the national interest test gave the treaty power a potentially unlimited reach. But the no-conflict test set outer limits beyond which not even the growth of national interest could carry the treaty power. In that sense, the no-conflict test was superior to the national interest test.

At the same time, the no-conflict test bolstered the national interest test within those limits, and this was its immediate role within the *Holland* opinion. Even if the states were competent to preserve migratory birds, said Holmes, "the question is [still] whether the United States is forbidden to act."[142] National interest in birds would not cease to afford constitutional support to the Migratory Bird Treaty even in the face of a showing that the states severally could protect the creatures. National interest became irrelevant only if there was a specific provision barring federal action. Thus emerges the fundamental importance of Holmes's review of the Tenth Amendment's relation to treaty adjudication. It aimed at determining whether the Amendment contained restrictions which would set limits on treaties under the no-conflict test.

V. Justice Holmes's Opinion: An Evaluation

Justice Holmes eschewed defending the Migratory Bird Treaty Act of 1918 as a proper regulation of interstate commerce or federal property. This is understandable. While such non-treaty arguments remained available despite their rejection by lower courts in *Shauver* and related cases and while they may have been persuasive to Holmes himself, it is doubtful that they could have carried a majority of the Supreme Court in 1920. Proponents of the 1913 bird protection law had themselves been uncertain of its constitutionality, but saw a treaty as offering a firm foundation. Moreover, the differing responses of the lower courts to the two

142. *Id.* at 435.

bird protection laws is instructive. The treaty did make a difference in judicial outcome. In *Holland* itself, the district court had agreed that the 1918 law would have been unconstitutional were it not enacted pursuant to a treaty. Finally, in its brief before the Supreme Court in *Holland,* the Government relegated its non-treaty arguments to an appendix, indicating what several well-placed and interested contemporaries thought would impress the Court.

In operative terms, Holmes approached the testing of treaties far differently than most of his judicial contemporaries would have approached the testing of ordinary legislation. The minimal requirement for any federal action was an express or implied constitutional authorization, and there were also relevant prohibitions to be accommodated. With legislation, both these considerations imposed real limitations. However, because the Constitution gave blanket authorization to the President to make treaties with the advice and consent of the Senate, the only substantial restriction on treaties for Holmes was that they not violate constitutional prohibitions. To be sure, they had to involve matters of national interest, but "national interest" was hardly a narrow category, particularly with Holmes's organic view of the Constitution.

Although Holmes did not review it in his opinion, there was solid judicial precedent for testing treaties by the no-conflict and national interest requirements, even in 1920.[143] The best-known

143. See, *e.g.,* New Orleans v. United States, 10 Peters 662, 736 (1836); Doe v. Braden, 16 How. 635, 657 (1853); The Cherokee Tobacco, 11 Wall. 616, 620-21 (1870); Holden v. Joy, 17 Wall. 211, 242 (1872); Geofroy v. Riggs, 133 U.S. 258, 266-67 (1890); Downes v. Bidwell, 182 U.S. 244, 294 (1901) (White, Shiras, and McKenna, JJ., concurring); *id.* at 370 (Fuller, C.J., Harlan, Brewer, and Peckham, JJ., dissenting). For non-judicial comment affirming this point during the two decades preceding, and also roughly contemporaneous with, the *Holland* case, see 1 BUTLER, THE TREATY-MAKING POWER OF THE UNITED STATES 349-64 (1902); 1 WILLOUGHBY, THE CONSTITUTIONAL LAW OF THE UNITED STATES 493-504 (1910); CORWIN, NATIONAL SUPREMACY: TREATY POWER VS. STATE POWER (1913); TUCKER, LIMITATIONS ON THE TREATY-MAKING POWER 332-40 (1915); SUTHERLAND, CONSTITUTIONAL POWER AND WORLD AFFAIRS 141-62 (1919); Wright, *The Constitutionality of Treaties,* 13 AM. J. INT'L L. 242, 252-60 (1919); Boyd, note 3 *supra,* at 414-21. Of these commentators, it is especially noteworthy that even Butler, who sought to document the sweeping scope of the treaty power, recognized limitations to it largely comparable to those stated by Justice Field in Geofroy v. Riggs, *supra,* as quoted in text *infra,* at note 144. He nevertheless found discussion of such limitations "necessarily academic"

formulation came from Justice Field, speaking for the Court in *Geofroy v. Riggs*:[144]

> The treaty power, as expressed in the Constitution, is in terms unlimited except by those restraints which are found in that instrument against the action of the government or of its departments, and those arising from the nature of the government itself and of that of the States. It would not be contended that it extends so far as to authorize what the Constitution forbids, or a change in the character of the government or in that of one of the States, or a cession of any portion of the territory of the latter, without its consent. . . . But with these exceptions, it is not perceived that there is any limit to the questions which can be adjusted touching any matter which is properly the subject of negotiation with a foreign country.

As Professor Henkin has observed, this statement and those like it "assert[ed] the fullness of [the treaty] power rather than restrictions upon it."[145] With the exception of the restriction, probably outmoded by 1920,[146] on ceding state territory, Field specified no well-defined limitations other than the requirement that a treaty not "authorize what the Constitution forbids." And as regards the supremacy of treaties over state constitutions and laws, this had received judicial recognition as early as 1796 in *Ware v. Hylton*,[147] as well as in subsequent cases.[148] (In this instance, Holmes did review the precedents.)[149] In short, Holmes's opinion hardly turned on novel judicial doctrine. Indeed, his approach had been antici-

---

and "practically of little value." 1 BUTLER, *supra*, at 363. For background to the specific issue presented in *Holland*, the best account is CORWIN, *supra*. For recent commentary on the constitutional status of treaties, see HENKIN 137-56, 383-404 nn.33-92.

144. 133 U.S. 258, 267 (1890) (dictum).

145. HENKIN 141.

146. CORWIN, note 143 *supra*, at 133-34, 190-91.

147. 3 Dallas 199 (1796).

148. See cases cited 252 U.S. at 434-35. In 1913, Professor Corwin was able to find only one federal case after the early 1880s in which a state's so-called reserved powers were recognized, and then only in dictum, as a limitation on the federal treaty power. That case was Cantini v. Tillman, 54 Fed. 969, 976 (C.C.D. S.C. 1893). See CORWIN, note 143 *supra*, at 195.

149. 252 U.S. at 434-35.

pated by the lower court decisions favorable to the 1918 law,[150] and even by that guardian of the Tenth Amendment, Justice Brewer, some eleven years earlier.[151]

To go beyond an examination of judicial precedent and ask whether *Holland* accorded with the understandings of those who framed and ratified the Constitution might seem to impose an unfair standard. On quick reading, Holmes appeared to rule out recourse to the original understanding when he advanced his organic view of the Constitution. In fact, however, he only said that *Holland* should "be considered . . . *not merely* in [light] of what was said a hundred years ago."[152] In context, he was further arguing not that constitutional tests changed over time, but that the concept of national interest, which was central to one of his constitutional tests, was expansible; and this, he implied at least, was realized by those who called the nation into life.[153] So an examination of the original understanding of treaties is in order.

Consistent with the "plain words" of the document, the consensus in 1787-88 of Federalists and Antifederalists alike was that the Constitution made treaties supreme over state constitutions and laws.[154] In fact, one of the complaints which led to the Fed-

150. See cases cited in notes 80 and 87 *supra.*

151. Keller v. United States, 213 U.S. 138, 147 (1909). Here the Court, per Justice Brewer, held unconstitutional, as beyond the delegated powers of the federal government, a statutory provision that prohibited the willful and knowing harboring of a female alien within three years of her arrival in the United States for purposes of prostitution. In reviewing possible grounds for upholding the provision, Justice Brewer commented: "By § 2 of Article II of the Constitution, power is given to the President, by and with the advice and consent of the Senate, to make treaties, but there is no suggestion in the record or in the briefs of a treaty with the King of Hungary [of whom the harbored alien was a subject] under which this legislation can be supported." 213 U.S. at 147. Brewer, it should be noted, had written the opinion of the Court in Kansas v. Colorado, 206 U.S. 46 (1907), on which opponents of both the 1913 and the 1918 bird protection laws placed so much reliance. Hence the implication of his remark in *Keller* seems particularly significant in illustrating the unexceptional nature of the doctrine Holmes advanced in *Holland.*

152. 252 U.S. at 433. (Emphasis added.)

153. *Ibid.*

154. THE FEDERALIST, No. 64, at 436-37 (Cooke ed., 1961) (John Jay); [James Iredell], *Answers to Mr. Mason's Objections to the New Constitution . . . ,* in PAMPHLETS ON THE CONSTITUTION OF THE UNITED STATES . . . 1787-1788, at 355-56 (Ford ed., 1888) [hereinafter cited as FORD, PAMPHLETS], [David Ramsey], *An Address to the Freemen of South Carolina . . . , id.* at 376; *Letters of Luther Martin, IV,* in ESSAYS ON THE CONSTITUTION OF THE UNITED STATES . . .

eral Convention of 1787 was the inability of the Confederation government to obtain state compliance with treaty provisions.[155] And *Ware v. Hylton,* besides providing strictly judicial precedent, further evidences the understanding of those close in time to the period of framing and ratification.[156]

The more difficult task is to determine whether Holmes's view of the relation of treaties to the federal Constitution is historically correct. Here, in questioning whether treaties need meet any test other than being properly made, approved, and ratified, Holmes at least superficially ignored the conventional explanation for the wording of the Supremacy Clause. That explanation stresses that the clause was dictated not by a desire on the founders' part to place treaties above the Constitution but by the need to ensure continued validity for treaties made prior to the adoption of the Constitution.[157] Professor Henkin opines that "[p]erhaps Holmes

---

1787-1788, at 361 (Ford ed., 1892); [Samuel Bryan (?)], *Letters of Centinel,* in PENNSYLVANIA AND THE FEDERAL CONSTITUTION, 1787-1788, at 580, 582, 610-11 (McMaster & Stone eds., 1888); *Petition of Group Chaired by Blair M'Clenachan to Pennsylvania General Assembly, id.* at 564; *Reasons of Dissent of the Minority,* id. at 476; 2 ELLIOT, DEBATES IN THE SEVERAL STATE CONVENTIONS ON THE ADOPTION OF THE FEDERAL CONSTITUTION . . . IN 1787 . . . , at 506-07 (2d ed. 1888) (James Wilson, Pennsylvania) [hereinafter cited as ELLIOT]; 3 *id.* at 500-14 (various speakers, Virginia). See generally 1 BUTLER, note 143 *supra,* at 341-92; Corwin, note 143 *supra,* at 66-74; CRANDALL, TREATIES: THEIR MAKING AND ENFORCEMENT 53-63 (2d ed. 1916); COWLES, TREATIES AND CONSTITUTIONAL LAW: PROPERTY INTERFERENCES AND DUE PROCESS OF LAW 18-19 (1941).

155. *E.g.,* Farrand, *The Federal Constitution and the Defects of the Confederation,* 2 AM. POL. SCI. REV. 532, 535-36 (1908); see also MARKS, INDEPENDENCE ON TRIAL: FOREIGN AFFAIRS AND THE MAKING OF THE CONSTITUTION 3-15, 151-52 (1973).

156. "[T]he contemporaries of the constitution have claims to our deference . . . because they had the best opportunities of informing themselves of the understanding of the framers of the constitution, and of the sense put upon it by the people, when it was adopted by them. . . ." Ogden v. Saunders, 12 Wheat. 212, 290 (1827) (Johnson, J., separate opinion). The same consideration is pertinent to the following discussion of the ratification debates of 1787-88, the Judiciary Act of 1789, and the Jay Treaty debate.

157. See, *e.g.,* Reid v. Covert, 354 U.S. 1, 16-17 (1957) (Black, J., plurality opinion); RAWLE, A VIEW OF THE CONSTITUTION OF THE UNITED STATES OF AMERICA 66-67 (2d ed. 1829); CORWIN, THE PRESIDENT: OFFICE AND POWERS, 1787-1957, at 421 n.17 (1957); SCHWARTZ, A COMMENTARY ON THE CONSTITUTION OF THE UNITED STATES: THE POWERS OF GOVERNMENT 135-36 (1963); citations in note 97 *supra. Cf.* 1 BUTLER, note 143 *supra,* at 321; CORWIN, note 143 *supra,* at 64; CRANDALL, note 154 *supra,* at 50.

did not know of that suggestion; perhaps he did not accept it."[158] The first possibility is doubtful. Missouri and Kansas advanced the explanation in their briefs before the Court. It is more likely that Holmes either did not accept the explanation or, accepting it, still concluded that treaties were to be held to tests different in practice from those applied to ordinary laws. Which of these alternatives is correct cannot be determined conclusively. The second is consistent with the history that Holmes probably knew well,[159] and with the fact that he ultimately did offer constitutional tests for treaties, whereas rejection of the conventional explanation could tend to endow treaties with an extra-constitutional status.

The direct evidence from 1787 in support of the conventional explanation for the wording regarding treaties in the Supremacy Clause, and thus in support of the subordination of treaties to the Constitution, is slim. The phrase "authority of the United States" appeared early in the Philadelphia Convention, on 31 May 1787.[160] Finally, on 23 August, the Convention approved the Supremacy Clause with wording that "all Treaties made under the authority of the U.S. shall be the supreme law."[161] On 25 August, the clause "was reconsidered and after the words 'all treaties made,' were inserted . . . the words 'or which shall be made [.]' This insertion was meant to obviate all doubt concerning the force of treaties preexisting, by making the words 'all treaties made' to refer to them,

158. HENKIN, 138.

159. Consider for example, what Holmes must have reviewed in editing Kent's *Commentaries,* which include this footnote: "The treaty-making power is necessarily and obviously subordinate to the fundamental laws and constitution of the state, and it cannot change the form of the government, or annihilate its constitutional powers." KENT, COMMENTARIES ON AMERICAN LAW *287 n.(a) (Holmes ed. 1873) (quoting Joseph Story). See generally HOWE, JUSTICE OLIVER WENDELL HOLMES: THE PROVING YEARS, 1870-1922, at 12-23 (1963). Holmes was familiar, too, with the Judiciary Act of 1789, which in § 25 contained implications for the status of treaties under the Constitution.

160. See 1 FARRAND, THE RECORDS OF THE FEDERAL CONVENTION OF 1787, at 47 (rev. ed. 1937) (amendment by Benjamin Franklin, 31 May 1787, to the "Virginia Plan") [hereinafter cited as FARRAND]. The phrase also appeared in the provision of the initial draft of the "New Jersey Plan" which evolved into the Supremacy Clause of the finished Constitution. See *id.* at 245 (15 June 1787). The history of the phrase is conveniently summarized in Myers, *Treaty and Law under the Constitution,* 26 DEPT. STATE BULL. 371, 373-76 (1952).

161. 2 FARRAND, 389.

as the words inserted would refer to further treaties."[162] This passage from Madison's notes hints—but only hints—that "the authority of the United States" was designed to encompass both past and future treaties. It also reveals that the phrase itself did not have a precise enough intension to satisfy the delegates. Aside from this, there is no direct evidence one way or another in the extant records of the Convention to illuminate the status of treaties vis-à-vis the Constitution. What seems likely is that, had the delegates firmly sensed the wording of the Supremacy Clause freed treaties from all constitutional controls save the requirements regarding their making and approval, then someone would have commented how the clause transgressed American notions of limited government.

Somewhat better evidence comes from the state ratification contests in 1787-88. Here, admittedly, the main concern was the impact treaties would have on state constitutions and laws. As already noted, Federalists and Antifederalists agreed that, consistent with the plain wording of the document, the clause did in fact establish the supremacy of treaties over state constitutions and laws.[163] Even the Antifederalist Federal Farmer, who observed that "[i]t is not said that these treaties shall be made in pursuance of the constitution—nor are there any constitutional bounds set to those who shall make them,"[164] most feared their impact on state law.

162. *Id.* at 417. Professor Henkin correctly observes that the element of the Supremacy Clause which is crucial to its encompassing preexisting treaties is the phrase "made, or which shall be made." HENKIN, 138, 385 n.38. If, however, he means to suggest that retention of "the Authority of the United States" was therefore designed to accomplish other purposes (his text is not entirely clear about his position), I would disagree. Clarity, rather than a studied avoidance of all redundancy, emerges as the object in reformulating the clause on 25 August. See 2 FARRAND, 417; McLaughlin, *The Scope of the Treaty Power in the United States, I,* 42 MINN. L. REV. 709, 730-31 (1958).

163. See citations in note 154 *supra.* For an example, apropos *Holland,* of what I believe to be an incorrect reading of the Antifederalists' concerns respecting treaties and the Supremacy Clause (that is, that they were fearful the Supremacy Clause made treaties supreme over the *federal* Constitution), see Deutsch, *Treaty-Making Clause: A Decision for the People of America,* 37 A.B.A.J. 659, 662 (1951).

164. *Letters from the Federal Farmer to the Republican* in FORD PAMPHLETS, at 312. The "Federal Farmer," formerly thought to have been Richard Henry Lee, now appears unidentifiable. See Wood, *The Authorship of the* Letters from the Federal Farmer, 31 WM. & MARY Q. 299 (3d ser. 1974).

In context, in fact, the "constitutional bounds" he was concerned with are the bounds necessary to preserve state constitutions and bills of rights. Still, the Federal Farmer's comment gives an inkling that he saw treaties as paramount to the United States Constitution.[165] The same suggestion emerges from an amendment proposed in Pennsylvania that "neither shall any treaties be valid which are in contradiction to the Constitution of the United States."[166] But the hint is again weak. The amendment may have reflected anxiety that treaties would have an extra-constitutional status, or it may simply have reflected the common Antifederalist desire to clarify the Constitution.

Running counter to such fears, and presumably better indicators of the Constitution's design, are several statements by Federalists in the Virginia ratifying convention, where the most extensive debate on treaties occurred. Responding to charges from Patrick Henry, the Federalists pointed primarily to the political check on treaties which the two-thirds rule in the Senate provided.[167] Three Federalists went further, however. James Madison denied the treaty power carried the authority to dismember the nation.[168] He also argued that "exercise of the power must be consistent with the object of the delegation," which, in the case of treaties, was "the regulation of intercourse with foreign nations, and is external."[169] Citing the provision of Article IV that "nothing in this Constitution shall be so construed as to prejudice any claims of the United States, or of any particular state," Governor Edmund Randolph concurred that the power carried no potential of dismemberment.[170] Moreover, the President and Senate, being subordinate to the Constitution whose creatures they were, could not by treaty alter the functions of the various departments of government.[171] The most restrictive reading came from George Nicholas, who in-

165. See FORD PAMPHLETS, at 312-13. *Cf.* 4 ELLIOT, 215 (William Lancaster, North Carolina).

166. PENNSYLVANIA AND THE FEDERAL CONSTITUTION, note 154 *supra,* at 463.

167. ELLIOT, 499-516. The issue also received less focused attention throughout the Virginia convention.

168. *Id.* at 500, 514.

169. *Id.* at 514.

170. *Id.* at 504.

171. *Ibid.*

terpreted the requirement that treaties be under "authority of the United States" as meaning "no treaty . . . shall be repugnant to the spirit of the Constitution, or inconsistent with the delegated powers."[172]

Those Virginia Federalists who discussed the issue thus perceived constitutional limits to treaties beyond the minimal requirement that they be properly made. That they failed to enumerate these limits in detail is not surprising. The problem which set the context for the Virginia debate was the narrow one of the threatened closing of the Mississippi River,[173] which explains the special concern about dismemberment of the nation. Furthermore, in the absence of a well-developed notion that the courts would exercise the power of constitutional review,[174] the question of well-defined limitations, such as might be applied by courts, was unlikely to arise. And, as Madison indicated, future developments would influence what objects might come within the scope of the treaty power.[175] In short, the occasion did not demand close attention to specific limitations. What does emerge, however, is (1) the view that treaties would be restricted to foreign objects (but with recognition that this was an expansive category), and (2) the implication, arising in various ways, that treaties could not infringe constitutional guarantees and prohibitions.

Consistent with these comments from the Virginia convention, the Judiciary Act of 1789 also hints that Americans originally concluded that treaties would be subject to constitutional tests. In its

172. *Id.* at 507.

173. See *e.g., id.* at 500-02; MARKS, note 155 *supra,* at 197-98; Warren, *The Mississippi River and the Treaty Clause of the Constitution,* 2 GEO. WASH. L. REV. 271, 296-98 (1934); see also Bestor, *Separation of Powers in the Domain of Foreign Affairs: The Intent of the Constitution Historically Examined,* 5 SETON HALL L. REV. 527, 613-60 (1974).

174. Although debate continues over the original understanding of judicial review, Professor Levy has aptly summed up the situation: "The precedents [as of 1787-88] tend not to show that the courts could pass on the constitutionality of the general powers of the legislatures." Levy, *Judicial Review, History, and Democracy: An Introduction,* in JUDICIAL REVIEW AND THE SUPREME COURT 1, 11 (1967). Hence, whatever germs of judicial review were present in the Constitution and in American constitutional thought at this time, I think it safe to conclude that the concept was not sufficiently developed so as to force contemporaries to face comprehensively the question of how courts would test treaties.

175. 3 ELLIOT, 514-15.

crucial § 25, the act provided for appeals to the Supreme Court by writ of error when the highest court of a state had questioned the validity of a treaty.[176] This, unfortunately, indicates nothing about what specific tests the framers of the act thought might be applied to treaties.

The fullest indication of early American conclusions about the constitutional status of treaties comes from a largely neglected aspect of the Jay Treaty debate in the House of Representatives in March and April of 1796.[177] The immediate issue in the debate was whether the House might, consistent with the Constitution's treaty provisions, pass independent judgment on the merits of a treaty in the course of appropriating funds needed to implement it, and thus whether the House could call on the President to provide papers relating to the negotiation of the treaty. (This is the aspect of the debate which has attracted the main attention of constitutional scholars.)[178] Nonetheless, the constitutional status of treaties also received comment, for the Federalists conceded that the House could pass independent judgment on treaties, but in only two instances: (1) where the treaty was, on its face, patently contrary to the nation's interests (in which case impeachment of the President was a proper recourse), and (2) where it was contrary to the Constitution.[179]

Of interest here is the latter concession, although like the Judiciary Act of 1789 it does not by itself reveal what tests the Federalists were willing to apply to treaties. Abstractly, they had a

176. 1 STAT. 73, 85-87, § 25 (1789).

177. The only study I have found which gives extended attention to the debate's implications for the early understanding regarding constitutional limitations on treaties, as opposed to its implications for the House's role in the treaty process, is BYRD, TREATIES AND EXECUTIVE AGREEMENTS IN THE UNITED STATES 35-59 (1960). My conclusions from the debate largely parallel Byrd's. The subject is also briefly discussed in MCLAUGHLIN, A CONSTITUTIONAL HISTORY OF THE UNITED STATES 260-61 (1935).

178. See, *e.g.*, CORWIN, note 157 *supra*, at 182; HENKIN, 161-62; SCHLESINGER, THE IMPERIAL PRESIDENCY 16-17 (1973); BERGER, EXECUTIVE PRIVILEGE: A CONSTITUTIONAL MYTH 171-79 (1974).

179. For Federalist statements developing various aspects of this position, see 5 ANNALS, cols. 426-27, 430-35, 438-39, 462, 530-32, 551, 593-94, 609-10, 621, 660-62, 671, 684-703, 712, 715-16, 989, 1016, 1160, 1204 (1849) (7 March-27 April 1796). (There are two editions of the *Annals;* all citations herein are to the edition which carries the running head "History of Congress" on *each* page.)

range of options. They may only have been conceding that a treaty must be properly made and approved. At the other extreme, they may have been admitting the Republican contention that at least without implementing legislation, a treaty could not infringe on the legislative power shared by the House of Representatives.[180]

While most Federalists did not carefully elaborate what they conceived would constitute an unconstitutional treaty, they clearly rejected the Republican position, but stopped short of simply requiring proper form. The common assumption was given voice by Theophilus Bradbury of Massachusetts, who, after remarking that "no laws inconsistent with that [the Constitution] can be passed, either by the Treaty or Legislative power," said that "the only *other* checks" on the treaty power of the President were the requirement for Senate approval and the possibility of impeachment.[181] So the requirement that a treaty be properly made and approved was in addition to the requirement that it be consistent with the Constitution.

Two other Federalists were more specific. Republican James Madison had charged that the Federalist position would allow treaties to contravene specific prohibitions placed on Congress in the body of the Constitution and in the Bill of Rights.[182] James Hillhouse of Connecticut denied the accusation, commenting that "it is a sound rule of construction, that what is forbidden to be done by all branches of the government conjointly, cannot be done by one or more of them separately."[183] Daniel Buck of Vermont added detail:[184]

> Is it not agreed by all, that if a Treaty violates the Constitution, it is void in itself? Does not the Constitution particularly point out how the Legislature shall be formed; what shall be the qualifications of its members, and how they shall be elected? Does it not point out, with the same precision, how each other department of Government shall be constituted and organized; and does

180. See, *e.g., id.,* cols. 493-94 (Madison, 10 March 1796), 738-43 (Gallatin, 24 March 1796).

181. *Id.,* col. 551 (14 March 1796). (Emphasis added.)

182. *Id.,* col. 491 (10 March 1796).

183. *Id.,* col. 671 (22 March 1796).

184. *Id.,* cols. 715-16 (24 March 1796).

> it not mark out the powers and limits of each? Does it not guaranty to each State its republican form of Government; and is not the right of altering or creating anew the Constitution reserved to the People? Look to the Constitution, and it will be found that the PRESIDENT, Senate, and this House, cannot, with all their combined powers, interfere with the personal security, personal liberty, or private property of the people, unless in raising taxes, and the mode in which this is to be done is directed by the Constitution.

Significantly, Republican William Giles of Virginia had already admitted the Federalists were not contending for an unlimited treaty power, but were qualifying it to the extent that "it is not to be supreme over the head of the Constitution."[185]

Besides agreeing that treaties must not contravene specific prohibitions of the Constitution, Federalists accepted another limit. James Hillhouse put it that a treaty must relate "to objects within the province of the Treaty-making power, a power which is not unlimited. The objects upon which it can operate are understood and well defined." "A treaty," he said, "is a compact entered into by two independent nations, for mutual advantage."[186] The same position was advanced by Theodore Sedgwick of Massachusetts. Of all those Federalists who spoke in the debate, Sedgwick was probably the most extreme advocate of a broad treaty power. For instance, while not actually contradicting his fellows on the point, he did not clearly admit that treaties could not contravene the Constitution.[187] But even he proposed a limited subject matter for treaties when he stated:[188]

> The power of treating between independent nations might be classed under the following heads: 1. To compose and adjust differences

185. *Id.,* col. 506 (10 March 1796). He added: "[B]ut in every other respect they contend that it shall be unlimited, supreme, undefined." In context, this referred to the Federalist claim that treaties might properly involve the usual objects of international negotiation and that the House had no authority to pass further judgment on their merits after Senate approval had been given. *Id.,* cols. 505-07.

186. *Id.,* cols. 660, 662 (22 March 1796).

187. *Id.,* cols. 514-30 (11 March 1796).

188. *Id.,* col. 516.

whether to terminate or to prevent war. 2. To form contracts for mutual security or defence; or to make Treaties, offensive or defensive. 3. To regulate an intercourse for mutual benefit, or to form Treaties of Commerce.

What emerges, then, from the Jay Treaty debate is a consensus. However they differed about the propriety of the House's calling on the President for papers and about the need to engage the House when treaties dealt with subjects ordinarily falling within the legislative realm, Federalists and Republicans saw treaties as subordinate, at minimum, to the Constitution. They agreed that treaties could not contravene its prohibitions and guarantees and could properly encompass only those objects of mutual interest between nations.[189]

Consistent with this consensus, and apropos *Missouri v. Holland,* the Jay Treaty debate also provides the earliest explicit instance I have found of what has become the common explanation of the wording of the Supremacy Clause. Several Federalists explained that the phrase "authority of the United States" was simply a convenient means of encompassing preexisting treaties made by the Continental and Confederation Congresses as well as those made under the Constitution by the President with the approval of the Senate.[190]

189. Accordingly, it is not surprising to find that a 1796 statute regulating the conveyance of Indian lands used the wording "treaty, or convention, entered into *pursuant* to the constitution." 1 STAT. 469, 472. (Emphasis added.) On the other hand, the Louisiana Purchase Treaty gave France and Spain commercial advantages in ports of Louisiana which they did not enjoy in other ports of the United States, and the treaty correspondingly made these Louisiana ports more attractive than others to French and Spanish shipping. The advantages continued for a period after Louisiana obtained statehood, in apparent violation of the clause of Art. 1, § 9, that "[n]o Preference shall be given by any Regulation of Commerce or Revenue to the Ports of one State over that of another." Farrand, *The Commercial Privileges of the Treaty of 1803,* 7 AM. HIST. REV. 494 (1902).

190. ANNALS, cols. 548 n.† (Bradbury, 14 March 1796), 669 (Hillhouse, 22 March 1796), 721 (Goodrich, 24 March 1796). *Cf. id.,* col. 558 (Page, 15 March 1796). One Republican made explicit the more restrictive reading that was implicit in many other Republican comments: that the "authority of the United States," under which treaties were to be made, comprised not only the Constitution, but also all existing laws. Hence a treaty contrary to existing law would be invalid. *Id.,* col. 578 (Brant, 15 March 1796). *Cf. id.,* cols. 592-93 (Findley, 16 March 1796).

The difficulty with Holmes's opinion therefore consists not in its reliance on the treaty power, nor, in the main, in the judicial and historical warrants underlying the tests he proposed for treaties. Instead, the problem is this: Holmes raised the question whether there were any limits to treaties so as to hint that there were none, and he then buried his actual answer–that is, his proposed limitations–in passages which lacked precision and were far more striking for their general constitutional ideas and aphorisms. As a result, he practically encouraged misinterpretation of the Court's decision. Thus, at one extreme, citing and quoting *Missouri v. Holland,* a California judge commented in 1937 that "[u]nder the present state of the law it may be conceded that it is uncertain whether there is any limitation at all on the treaty-making power of the federal government."[191] At the other, a congressman alleged in 1974 that the decision supported the proposition "that one . . . limit [to the treaty-making power] is that what an act of Congress could not do unaided, in derogation of the power of the Constitution, a treaty cannot do."[192]

Now state judges and federal congressmen, while bound to uphold the Constitution, are not always the most astute commentators on it. Yet the intense controversy which raged over *Missouri v. Holland* in the 1950s, during the Bricker Amendment debate, reveals a similar picture. Not only congressmen, who might be accused of political perversity in the episode, but also legal scholars were able to read quite different meanings into the *Migratory Bird* opinion. And some thirty years earlier, even so highly regarded a constitutionalist as Thomas Reed Powell could say of the opinion, in evident disregard of certain of Holmes's comments, that "[i]ts hint that there may be no other test to be applied than whether the treaty has been duly concluded indicates that the court might hold that specific constitutional limitations in favor of individual

191. Tokaji v. State Bd. of Equalization, 20 Cal. App.2d 612, 617 (1937). See United States v. Reid, 73 F.2d 153, 155 (9th Cir. 1934). Quite inexplicably, the quoted statement from *Tokaji* was followed by a quotation from *Holland* which included Holmes's comment, "We do not mean to imply that there are no qualifications to the treaty-making power; but they must be ascertained in a different way." 20 Cal. App.2d at 617.

192. 120 Cong. Rec. at S.465 (daily ed. 1974) (Sen. William L. Scott, 28 Jan. 1974, discussing the proposed American adherence to the Genocide Convention).

liberty and property are not applicable to deprivations wrought by treaties."[193] In fine, the opinion fits the pattern suggested by Justice Frankfurter, surely an admirer of Holmes, when he wrote that Holmes "spoke for the Court, in most instances tersely and often cryptically."[194] Similarly, Justice Brandeis, Holmes's colleague on the bench, might have had *Holland* in mind in remarking that Holmes did not "sufficiently consider the need of others to understand."[195]

VI. Holland's Impact and Latter-Day Significance

Did *Missouri v. Holland* have an impact on the law of the Constitution? At one level, obviously yes. In the 1930s and early 1940s, to be sure, courts evidenced a willingness to uphold conservation activities under the Commerce Clause, independent of a treaty base.[196] But in the doctrinal climate of the 1920s and early

193. Powell, *supra* note 3, at 13. Admittedly Professor Powell's next sentence read: "It would be going a step further to extend the same immunity to legislation enforcing treaties." *Id.* But I fail to see that this addition substantially rectified his interpretation of the limits Holmes placed on treaties. If anything, it compounded the error by ignoring Holmes's evident willingness to judge treaties and legislation implementing them by the same standards. See text *supra,* at note 119.

194. Federal Maritime Board v. Isbrandtsen Co., 356 U.S. 481, 523 (1958) (dissenting).

195. Quoted in Bickel, The Unpublished Opinions of Mr. Justice Brandeis 226-27 (1957). Regarding Holmes's cryptic style, see Rogat, *Mr. Justice Holmes: A Dissenting Opinion,* 15 Stan. L. Rev. 3, 9-10 & nn.26-33 (1962).

196. Bogle v. White, 61 F.2d 930, 931 (5th Cir. 1932); Cerritos Gun Club v. Hall, 96 F.2d 620, 623-28 (9th Cir. 1938); Bailey v. Holland, 126 F.2d 317, 321 (4th Cir. 1942). Interestingly, the Court of Appeals in the *Cerritos Gun Club* case took issue with Holmes's contention in *Holland* that title to migratory birds was a dubious concept. Said the Court: "Naturalists and the public at large have learned much about the habits and actual and possible domestication of these migratory fowl since Mr. Justice Holmes wrote [his] opinion." 96 F.2d at 624. As items of property, the birds were thus articles in commerce. *Id.* at 623-27. But *cf.* Toomer v. Witsell, 334 U.S. 385, 401 (1948); Takahashi v. Fish and Game Commission, 334 U.S. 410, 421 (1948) (both using *Holland* to support contention that fish are not in the possession of anyone in the ordinary property sense); Russo v. Reed, 93 F. Supp. 554, 560 (D. Me. 1950); Koop v. United States, 296 F.2d 53, 59 (8th Cir. 1961) (both using *Holland* to support contention that state ownership of wildlife is doubtful).

1930s, the decision in *Holland* did facilitate federal conservation policy and thus influenced not only law but also America's natural environment. It established the validity of the 1918 Migratory Bird Treaty Act,"[197] thereby "first plac[ing] this new policy [of bird protection] on a firm foundation."[198] It also contributed a constitutional base for such further conservation efforts as federal wildlife preserves, reforestation projects, and associated state land donations.[199]

At another level, the case's impact is more difficult to assess. Precisely because it accorded with judicial precedent as to the supremacy of treaties over state law, and because it recognized, however cryptically, the commonly accepted limits to treaties, it hardly staked out new ground. It nevertheless fulfilled the prophecy of one commentator ten years earlier "that the *obiter* doctrine that the reserved rights of the States may never be infringed upon by the treaty-making power will sooner or later be frankly repudiated by the Supreme Court."[200] Accordingly, it served as convenient precedent in later cases which drew into issue the supremacy of treaties or executive agreements over state law.[201] Meanwhile, it also bolstered dicta that treaties could not authorize what the

197. For cases citing *Holland* in this connection, see Shouse v. Moore, 11 F. Supp. 784, 785 (E.D. Ky. 1935); Cerritos Gun Club v. Hall, 21 F. Supp. 163, 165 (S.D. Cal. 1936), *aff'd* 96 F.2d 620 (9th Cir. 1938); Cochrane v. United States, 92 F.2d 623, 626-27 (7th Cir. 1937); United States v. Reese, 27 F. Supp. 833, 836 (W.D. Tenn. 1939); Lansden v. Hart, 180 F.2d 679, 683 (7th Cir. 1950); Bishop v. United States, 126 F. Supp. 449, 450-51 (Ct. Cl. 1954); Converse v. United States, 227 F.2d 749, 750 (6th Cir. 1955).

198. United States v. Otley, 34 F. Supp. 182, 191 (D. Ore. 1940).

199. United States v. 2,271.29 Acres, 31 F.2d 617, 621 (W.D. Wis. 1928); United States v. 546.03 Acres, 22 F. Supp. 775, 777 (W.D. Pa. 1938); *In re* United States, 28 F. Supp. 758, 763-64 (W.D. N.Y. 1939); Swan Lake Hunting Club v. United States, 381 F.2d 238, 242 (5th Cir. 1967).

200. 1 WILLOUGHBY, note 143 *supra,* at 503. Whether Willoughby himself thought *Holland* fulfilled his prediction is unclear, for he included the same statement at page 569 in his 1929 edition.

201. Asakura v. Seattle, 265 U.S. 332, 341 (1924); Santovincenzo v. Egan, 284 U.S. 30, 40 (1931); United States v. Belmont, 301 U.S. 324, 331-32 (1937); Amaya v. Stanolind Oil and Gas Co., 158 F.2d 554, 556 (5th Cir. 1947); Power Authority of New York v. F.P.C., 247 F.2d 538, 545 (D.C. Cir. 1957) (Bastion, J., dissenting). *Cf.* United States v. California, 332 U.S. 19, 45 (1947) (Frankfurter, J., dissenting); Sohappy v. Smith, 302 F. Supp. 899, 912 (D. Ore. 1969).

Constitution forbade and must involve matters of national interest.[202] Finally, in 1957, Justice Black explained that "nothing in *Missouri v. Holland* . . . is contrary to the position taken here [that treaties must not contravene constitutional guarantees]. There the Court carefully noted that the treaty involved was not inconsistent with any specific provision of the Constitution."[203]

More specifically, Holmes's passage about an organic constitution that takes shape from, and grows with, the nation's history and needs[204] has proved eminently quotable and serviceable. In the *Mortgage Moratorium Case,*[205] this comment, along with Chief Justice Marshall's in *McCulloch v. Maryland* about "a constitution intended to endure for ages to come,"[206] aided Chief Justice Hughes in demonstrating that "[i]t is no answer to say that this public need [for mortgage payment relief] was not apprehended a century ago, or to insist that what the provision of the Constitution [concerning the obligation of contracts] meant to the vision of that day it must mean to the vision of our time."[207] The same passage from Holmes's opinion has been enlisted to support broadened individual protection in the areas of denaturalization,[208] racial discrimination,[209] and

202. Asakura v. Seattle, 265 U.S. 332, 341 (1924); Amaya v. Stanolind Oil and Gas Co., 158 F.2d 554, 556 (5th Cir. 1947); Power Authority v. Federal Power Commission, 247 F.2d 538, 542 (D.C. Cir. 1957); Pierre v. Eastern Air Lines, 152 F. Supp. 486, 488 (D.N.J. 1957).

203. Reid v. Covert, 354 U.S. 1, 18 (1957) (plurality opinion).

204. 252 U.S. at 433.

205. Home Building and Loan Assoc. v. Blaisdell, 290 U.S. 398 (1934).

206. McCulloch v. Maryland, 4 Wheat. 316, 407 (1819), quoted more exclusively in 290 U.S. at 443.

207. 290 U.S. at 443. For quotations, in turn, of Hughes's quoting Holmes, see, *e.g.*, Campbell v. Alleghany Corp., 75 F.2d 947, 955 (4th Cir. 1935) (upholding Federal Bankruptcy Act of 1934); Gomillion v. Lightfoot, 270 F.2d 594, 604 (5th Cir. 1959) (Brown, J., dissenting, and contending that a gerrymander had occurred, contrary to the spirit of the Constitution). *Rev'd* 364 U.S. 339 (1960) (held to be violation of Fifteenth Amendment). Intriguingly, four years before the *Mortgage Moratorium* case, a federal district judge used the same passage from *Holland* that Hughes quoted, but for the opposite purpose of restricting governmental power. Simply because the Constitution specified an amendment process, the judge argued, its terms could not always be followed literally, for the spirit of the document had to be considered. See United States v. Sprague, 44 F.2d 967, 981 (D.N.J. 1930) (holding the Prohibition Amendment unconstitutional), *rev'd* in 282 U.S. 716 (1931).

208. Baumgartner v. United States, 322 U.S. 665, 673 (1944).

209. Bell v. Maryland, 378 U.S. 226, 316-17 (1964) (Goldberg, J., concurring).

malapportionment of voting districts.[210] But *Holland* also carries the contrary potential as illustrated by Justice Frankfurter's use of it in *Dennis,* to the end that the federal government possessed authority, commensurate with national needs, to protect national security.[211] And again emerging on the side of governmental authority over the individual, the case has received mention in lower court dicta that even absent other constitutional grounds, federal narcotic drug legislation could be upheld as a necessary and proper means of implementing various drug treaties to which the United States is a party.[212]

Still, Professors Chafee, Sutherland, and Henkin have argued persuasively that the Court's post-1937 acquiescence in federal programs based on the taxation and commerce powers and on the Fourteenth Amendment has deprived *Missouri v. Holland* of much of its earlier significance. Legislation which in the 1920s and early 1930s might only have passed judicial scrutiny were it pursuant to a treaty now could easily stand alone.[213] The obvious example in-

210. Fortson v. Morris, 385 U.S. 231, 247-48 (1966) (Fortas, J, dissenting).

211. Dennis v. United States, 341 U.S. 494, 519-20 (1951) (Frankfurter, J., concurring). For other uses of *Holland* to support the proposition that where national action is required, the necessary power will likely be found to exist, see United States v. American Bond & Mortgage Co., 31 F.2d 448, 454 (N.D. Ill. 1929); *In re* United States, 28 F. Supp. 758, 763-64 (W.D. N.Y. 1939). *Cf.* Kentucky Whip and Collar Co. v. Illinois Central Railroad Co., 12 F. Supp. 37, 42 (W.D. 1935). Apropos the 1974 impeachment debate, a federal Court of Appeals judge, speaking off the bench, found *Holland* to support the notion that a President has considerable latitude in the exercise of his duties; hence for him to be impeached and convicted for exercising such discretion might provide grounds for Supreme Court review of his conviction. Gibbons, *The Interdependence of Legitimacy: An Introduction to the Meaning of Separation of Powers,* 5 Seton Hall L. Rev. 435, 485-86 n.218 (1974).

212. United States v. Eramdjian, 155 F. Supp. 914, 920 (S.D. Cal. 1957); United States v. Contrades, 196 F. Supp. 803, 812 (D. Hawaii 1961); United States v. Rodriguez-Camacho, 488 F.2d 1220, 1222 (9th Cir. 1972); United States v. LaFroscia, 354 F. Supp. 1338, 1341 (S.D. N.Y. 1973); United States v. Maiden, 355 F. Supp. 743, 749 n.5 (D. Conn. 1973). The suggestion appeared earlier, but without reference to *Holland,* in Stutz v. Bureau of Narcotics, 56 F. Supp. 810 (N.D. Cal. 1944).

213. Chafee, *Federal and State Powers under the U.N. Covenant on Human Rights,* 1951 Wis. L. Rev. 389, 400-24; Sutherland *Restricting the Treaty Power,* in A.A.L.S. Selected Essays of Constitutional Law, 1938-1962, at 160, 180 (1963); Henkin, note 79 *supra,* at 913-22; Henkin, 147.

volves regulation of child labor.[214] After the Court twice struck down child labor laws, in 1918 and 1922,[215] some proponents of regulation toyed with the treaty route, seeing no other alternative save the uncertain course of constitutional amendment.[216] (An amendment was sent to the states but received only twenty-eight ratifications.)[217] Then, in 1941, the Court upheld the New Deal's Fair Labor Standards Act, which included child labor provisions, on grounds that Congress had absolute control over commerce.[218] But, somewhat paradoxically, the *Migratory Bird Case* may itself have aided the judicial transformation which diminished its own importance, for Holmes's opinion arguably bolstered governmental activism.

Holland's faded significance did not prevent it from figuring in an attempted constitutional change which, had it succeeded, would have had consequences far beyond anything Holmes and his brethren in 1920 could have imagined they were setting afoot.[219] In 1952, Senator John Bricker introduced an Amendment to the Constitution to limit the treaty power.[220] As reported out of committee in June 1953, the Amendment provided, among other things, that

214. Wood, note 111 *supra, passim.* The drive to enact child labor legislation and to obtain its validation by the Supreme Court contains striking similarities to the movement for protection of migratory birds. The striking dissimilarity, of course, is the ultimate fate of the two efforts in the pre-1937 Court: the birds fared far better than the children, for reasons quite understandable within the doctrinal climate of the day.

215. Hammer v. Dagenhart, 247 U.S. 251 (1918); Bailey v. Drexel Furniture Co., 259 U.S. 20 (1922).

216. *See* Boyd, note 3 *supra,* at 429-31; Jackson, *The Tenth Amendment versus the Treaty-Making Power,* 14 Va. L. Rev. 331, 332 (1928) (attacking the validity of a child labor treaty); Note, 22 Mich. L. Rev. 457 (1924) (defending the validity of a child labor treaty).

217. Congressional Quarterly, Guide to the Congress of the United States 287 (1971).

218. United States v. Darby, 312 U.S. 100 (1941).

219. See, *e.g.,* Sutherland, note 213 *supra,* at 173-80.

220. 98 Cong. Rec. 899, 907-08 (1952). For background to the Amendment and the political dimensions of the fight over it, see Sutherland, note 213 *supra,* at 160-73; Schubert, *Politics and the Constitution: The Bricker Amendment during 1953,* 16 J. Pol. 257 (1954); Eisenhower, The White House Years: Mandate for Change, 1953-1956, at 340-48 (Signet ed. 1965); Parmet, Eisenhower and the American Crusades 305-12 (1972).

"[a] treaty shall become effective as internal law only through legislation which would be valid in the absence of a treaty."[221]

Although omitted from the version that came within one vote of Senate passage in February 1954,[222] this so-called "which" clause is a clue to part of the reasoning underlying the proposed Amendment. Bricker and others wished to eliminate what they saw as the Court-sanctioned route of amending the Constitution through treaty-making.[223] Highlighting the danger which the Amendment's backers thought needed remedying, Bricker characteristically told the Senate in January 1954 that Holmes had "suggested that 'under the authority of the United States' might mean 'nothing more than the formal acts required to make the convention.' " He further charged *Holland* had "repudiated" the early dictum "that the treaty power does not extend 'so far as to authorize what the Constitution forbids.' "[224] "One of the premises of that case [*Holland*]," he declared a week later, "is that Congress has unlimited

221. S. REP. No. 412, 83d Cong., 1st Sess. 1 (1953). Between 1952 and 1957, the Amendment appeared in several versions. These are reprinted and compared in *Hearings on S.J. Res. 3 before a Subcommittee of the Senate Committee on the Judiciary,* 85th Cong., 1st Sess., foldout facing 373 (1958).

222. 100 CONG. REC. 2358, 2374-75 (1954).

223. "This decision [*Holland*] in effect, and really for the first time, opened the way for amending the Constitution of the United States by and through a treaty, because it proclaims that an otherwise unconstitutional law may become constitutional when, as, and if the President negotiates a treaty on the subject and obtains approval of the Senate." Holman, *Treaty Law Making: A Blank Check for Writing a New Constitution,* 36 A.B.A.J. 707, 709 (1950). Holman, a former president of the American Bar Association, was one of the chief instigators of the Bricker Amendment and had been advocating such action for several years. See PARMET, note 220 *supra,* at 306-07. Three years later, another supporter explained: "The 'which clause' . . . would restrict implementing legislation to legislation valid in the absence of a treaty and is imperative because the case of *Missouri* v. *Holland* . . . made it perfectly clear that the Federal Government may, so long as that decision stands, invade and destroy all reserved powers of the states, arrogate to itself fields of legislative competence within the area of reserved powers where none existed without treaty, and regulate the purely internal concerns of the states, and the affairs of the citizens through the use of the treaty making power." Hatch, *The Treaty-Making Power: "An Extraordinary Power Liable to Abuse,"* 39 A.B.A.J. 808, 809 (1953).

224. 100 CONG. REC. 942-43 (28 Jan. 1954). For virtually identical comments from Bricker, see Bricker & Webb, *Treaty Laws vs. Domestic Constitutional Law,* 29 NOTRE DAME L. 529, 534-35 (1954).

power to legislate in implementation of a treaty."[225] Other supporters of the Amendment elaborated the need to correct the doctrines they alleged Holmes had advanced in legitimating a treaty made, as one article charged, "for the express purpose of conferring on [the federal government] legislative competence in domestic fields where it had none before."[226] For good reason an early student of the Bricker amendment debate labeled *Holland* "the *bête noire* of the [Amendment's] proponents."[227] Opponents correspondingly stressed that *Holland* did not place treaties and consequent legislation above the Constitution.[228]

225. 100 CONG. REC. 1333 (4 Feb. 1954). This remark would of course have been accurate had Bricker said "in implementation of a *valid* treaty."

226. Hatch, Finch & Ober, *The Treaty Power and the Constitution: The Case for Amendment,* 40 A.B.A.J. 207, 254 (1954). See Deutsch, *The Need for a Treaty Amendment: A Restatement and a Reply,* 38 A.B.A.J. 735 (1952); Finch, *The Need to Restrain the Treaty-Making Power of the United States within Constitutional Limits,* 48 AM. J. INT'L L. 57, 66-67 (1954); *Hearings on S.J. Res. 130 before a Subcommittee of the Senate Committee on the Judiciary,* 82d Cong., 2d Sess. 36-37 (1952) (statement of Alfred J. Schweppe, Chairman, Committee on Peace and Law through the United Nations of the American Bar Association) [hereinafter cited as *1952 Bricker Hearings*]; *Hearings on S.J. Res. 1 & 43 before a Subcommittee of the Senate Committee on the Judiciary,* 83d Cong., 1st Sess. 817 (1953) (testimony of Clarence Manion, former Dean, College of Law, University of Notre Dame) [hereinafter cited as *1953 Bricker Hearings*]; *Hearings on S.J. Res. 1 before a Subcommittee of the Senate Committee on the Judiciary,* 84th Cong., 1st Sess. 36-37 (1955) (comments of Sen. Dirksen) [hereinafter cited as *1955 Bricker Hearings*]; S. REP. No. 412, 83d Cong., 1st Sess. *supra* note 224, at 4, 6-7, 14, 16; 100 CONG. REC. 853-54, 1061-62, 1309-10, 1786-87, 2121, 2351 (comments of Sens. Butler of Md., Dirksen, Ferguson, Jenner, and Daniel, 27 Jan., 1, 4, 16, 23, 26 Feb. 1954). By 1955, however, at least some proponents of the Amendment were examining *Holland* more carefully. The majority committee report of that year stated: "Insofar as that decision relates to the first [that is, central] section of this amendment . . . the concern is not with the conclusion reached but rather with the expressions used in the course of the opinion." S. REP. No. 1716, 84th Cong., 2d Sess. 3 (1956).

227. Schubert, note 220 *supra,* at 286 n.121.

228. *E.g.,* Chafee, *Stop Being Terrified of Treaties: Stop Being Scared of the Constitution,* 38 A.B.A.J. 731, 732 (1952); Chafee, *Amending the Constitution to Cripple Treaties,* 12 LA. L. REV. 345, 354-56 (1952); Whitton & Fowler, *Bricker Amendment—Fallacies and Dangers,* 48 AM. J. INT'L L. 23, 30-40 (1954); Mathews, *The Constitutional Power of the President to Conclude International Agreements,* 64 YALE L.J. 345, 377 (1955); Corwin, *Memorandum on Senate Joint Resolution 1 as Reported by the Judiciary Committee, January 11, 1954,* printed in 100 CONG. REC. 859 (1954); *1952 Bricker Hearings* 78-80 (testimony of Theodore Pearson of Bar Association of the City of New York); *id.* at 413

The Bricker Amendment did not pass Congress, let alone enter the Constitution. At most, then, *Missouri v. Holland* served only to trigger and help fuel a national debate, although even in this regard its role needs qualifying. While the debate in legal periodicals and in the Senate contained frequent references to the case,[229] *Holland*'s importance pales amidst the complexities of historical causation. Reaction to several World War II agreements, alleged threats posed by the United Nations Human Rights and Genocide Conventions, and the frustrating "police action" in Korea all encouraged the movement for some kind of limitation on treaties and executive agreements.[230] In the absence of Holmes's holding that the Tenth Amendment was no bar to exercise of the treaty power and his easily misinterpreted remarks about other limitations on treaties, the legal and senatorial proponents of an Amendment might have lacked some of their ammunition. Also, the Amendment might have taken a different initial shape, perhaps without the "which" clause.[231] But such possibilities concerning an unsuc-

---

(memorandum submitted by Solicitor General Philip B. Perlman); *1953 Bricker Hearings* 905-06, 908, 910-12, 924-25 (statement of Attorney General Herbert Brownell); *1955 Bricker Hearings* 272-73 (statement of Bethuel M. Webster, former president of the Bar Association of the City of New York); S. REP. No. 412, note 221 *supra,* at 47 (minority views); S. REP. No. 1716, note 226 *supra,* at 26-27 (individual views of Sen. Kefauver); 100 CONG. REC. 658-59, 665, 1069-70, 2058-59 (Sen. Wiley, 22 Jan., 1, 19 Feb. 1954). See McLaughlin, *The Scope of the Treaty Power in the United States, II,* 43 MINN. L. REV. 651, 704-08 (1959).

229. See, *e.g.,* citations in notes 223-28 *supra.* In his memoirs, President Eisenhower also recognized the role of *Holland* in the Bricker Amendment controversy. See EISENHOWER, note 220 *supra,* at 340-41.

230. See, *e.g.,* Deutsch, note 163 *supra,* at 660-62; 98 CONG. REC. 907-14 (1952) (comments of Sen. Bricker and others on introduction of Bricker Amendment, 7 Feb. 1952); 100 *id.* at 934-37 (speech of Sen. McCarren, 28 Jan. 1954); Lofgren, *Mr. Truman's War: A Debate and Its Aftermath,* 31 REV. POL. 223, 236 (1969); PARMET, note 220 *supra,* at 308-09; THEOHARIS, THE YALTA MYTHS: AN ISSUE IN U.S. POLITICS, 1945-1955 105-06 & n.1, 131-33, 180-85 (1970).

231. For example, Alfred J. Schweppe, chairman of the American Bar Association's Committee on Peace and Law through the United Nations, testified in 1953 that the "which" clause "[was] intended specifically to limit the doctrine of *Missouri* v. *Holland.*" *1953 Bricker Amendment Hearings* 56. Similarly, the Majority Report on the Amendment in June 1953 indicated that *Holland* made necessary the "which" clause. S. REP. No. 412, note 221 *supra,* at 16. But the Report also indicated that the clause was "intended to correct the broad language in *U.S.* v. *Curtiss-Wright Export Corporation* [299 U.S. 304 (1936)]." *Id.* If the

cessful Amendment hardly suggest that *Holland* substantially influenced the actual course of events.

VII. Conclusion

This much is unexceptionable: The Supreme Court's doctrinal commitments, *circa* 1910-20, doubtless dictated that if the Court were to uphold federal migratory bird legislation, it would have to do so on grounds other than the federal government's authority over interstate commerce or federal property. Advocates of bird protection suspected this, perhaps as early as 1911. A treaty with Great Britain was subsequently concluded in 1916 and provided a base for new legislation in 1918 to replace the constitutionally questionable Migratory Bird Act of 1913. The strategy of the bird lovers, as well as that of the government in declining to push a test of the 1913 act, bore fruit in 1920. Justice Holmes, speaking for the Court, upheld the 1918 law as a necessary and proper means of implementing the Migratory Bird Treaty. The treaty itself, he argued, was a proper exercise of the treaty power, involving a subject of national concern and contravening no constitutional prohibitions.

Less certain are the reasons for the controversy surrounding *Missouri v. Holland.* Three considerations, however, seem important to its origins. First, the decision validated federal legislation which would arguably have failed had the Court ruled on it in its original form, devoid of a treaty base. This gave superficial plausibility to contention that for practical purposes the Court had held a treaty could amend the Constitution. Second, the result in *Holland* was put into bolder relief by virtue of the narrow play the "old" Court normally allowed to federal activity. Third, Holmes's opinion, while resting on grounds which were well established and historically warranted, failed to explicate those grounds fully and carefully and failed as well to clarify sufficiently the limits to the treaty power.

If these considerations were in fact crucial, then because of his

---

Amendment had passed Congress and then the states by narrow margins, it would be easier to assign a crucial influence to *Holland*.

cryptic remarks Holmes must bear some responsibility for the ensuing debate. Yet it should be remembered that he spoke with the approval of the Court. In a broader sense, the Court, but not so much Holmes, bears responsibility for helping to create the doctrinal climate which both forced reliance on the treaty power and put into sharp relief the result accomplished through that reliance.

What remains unexplained is why the controversy should have persisted into the 1950s, by which time the Court's commitments had markedly changed. Viewed objectively, the case now lacked importance as a potential source of domestic federal authority; the post-1937 Court had discerned new bases for expanded federal activity. In the new climate of concern over foreign involvement and enlarged federal authority, politics of course figured prominently in the renewed debate, with Holmes's passages making good targets. In part, too, the proponents of the Bricker Amendment may simply have overlooked how the Court's post-1937 flexibility rendered *Holland* largely redundant.[232] But something deeper may have been at work. The "Golden Age" of the judiciary for *Holland*'s critics in the 1950s was undoubtedly the period from the 1890s to 1937. Yet it was during this period that the Court had decided *Holland*. So, to speculate, reversal of the decision through constitutional amendment promised not only protection from "internationalist" schemes in the post–World War II era; it also promised to close off an avenue of encroachment on state authority and individual liberty which even a future right-headed Court might otherwise again follow.

A similar congeries of fears, frustrations, and yearnings could conceivably revive national interest in the case.[233] More likely, *Missouri v. Holland* will remain interred in the casebooks and history texts–and properly so. What is unlikely is that it will find crucial new applications in the development of the American Constitution.

232. Henkin, 147.

233. *Cf.* former Senator Bricker's plea in 1974: "[T]his matter [that is, the danger of treaties and executive agreements overriding the Constitution] is just as important now as it was twenty years ago. It is time for the public interest to be aroused—if the Constitution and the rights it guarantees are to be preserved." Bricker, *Bricker Amendment Still Apt Today,* Los Angeles Times, 10 Feb. 1974, § VI, p. 4, col. 5.

5

THE FOREIGN RELATIONS POWER: United States v. Curtiss-Wright Export Corporation: *An Historical Reassessment*

War between Paraguay and Bolivia broke out in June 1932.[1] Because both belligerents depended on outside military assistance, American armament manufacturers found the situation attractive, particularly in view of the depressed economy at home. But na-

1. For general historical background, see Divine, *The Case of the Smuggled Bombers,* in QUARRELS THAT HAVE SHAPED THE CONSTITUTION 210-21 (J. Garraty ed. 1966), on which I have relied for otherwise undocumented details in this and the following paragraph. For early analyses of *Curtiss-Wright,* see Levitan, *The Foreign Relations Power: An Analysis of Mr. Justice Sutherland's Theory,* 55 YALE L.J. 467 (1946); Patterson, *In re The United States v. The Curtiss-Wright Corporation* (pts. 1-2), 22 TEXAS L. REV. 286, 445 (1944); Quarles, *The Federal Government: As to Foreign Affairs, Are Its Powers Inherent as Distinguished from Delegated?,* 32 GEO. L.J. 375 (1944). These commentators are too ready to relegate parts of the opinion to the status of dicta, and they neglect the true implications of the "Story-Wilson" position. *See* pp. 182-95 *infra.* Recently, Raoul Berger has offered a more perceptive analysis of the historical evidence, although without, in my opinion, adequately confronting the dicta issue. *See* Berger, *War-Making by the President,* 121 U. PA. L. REV. 29, 69-75 (1972); Berger, *The Presidential Monopoly of Foreign Relations,* 71 MICH. L. REV. 1, 26-33 (1972). Berger's articles became available after the present article was largely completed; I am gratified that his studies, while less detailed specifically on *Curtiss-Wright,* generally support my own conclusions. For evi-

tional antiwar sentiment, revulsion at the slaughter in the Gran Chaco, and the urgings of Great Britain and the League of Nations prompted the American government to terminate the developing arms trade. On May 24, 1934, six days after it had been introduced, Congress approved a Joint Resolution providing that "if the President finds that the prohibition of the sale of arms and munitions of war in the United States to those countries engaged in conflict in the Chaco may contribute to the establishment of peace between those countries," he might proclaim an embargo on Amercian arms shipments to the belligerents. Violators would be fined, imprisoned, or both.[2] Franklin D. Roosevelt signed the Joint Resolution and issued an embargo proclamation on May 28.[3]

Curtiss-Wright Export Corporation, two associated companies, and four corporate officers were indicted in January 1936 for conspiring to sell aircraft machine guns to Bolivia in violation of the congressional resolution and presidential proclamation.[4] The defendants demurred to the indictment, arguing that the Joint Resolution unconstitutionally delegated legislative power to the Executive.[5] Agreeing, the district court ruled for the defendants.[6] This decision threatened the neutrality legislation which was evolving in response to European events; therefore the government appealed. Despite prior decisions evincing hostility to excessive delegations of power,[7] the Supreme Court in *United States v.*

---

dence of the continued legal relevance of the opinion, see, in addition to the statements cited in notes 12-30 *infra,* L. HENKIN, FOREIGN AFFAIRS AND THE CONSTITUTION 19-35 *passim* (1972). Henkin states that the opinion "remains authoritative doctrine." *Id.* at 25-26.

2. 48 Stat. 811 (1934). See note 33 *infra* for the complete text of the Joint Resolution.

3. 48 Stat. 1744 (1934).

4. Indictment, in Transcript of Record at 3, United States v. Curtiss-Wright Export Corp., 299 U.S. 304 (1936).

5. United States v. Curtiss-Wright Export Corp., 14 F. Supp. 230, 232 (S.D.N.Y. 1936) [*Curtiss-Wright* (*S.D.N.Y.*)]. Defendants also demurred on two grounds not relevant to this article: (1) Roosevelt's proclamation failed to meet the requirements set forth in the Joint Resolution, and (2) the defendants had been indicted after the President revoked his proclamation. The district court decided for the government on both points and the Supreme Court affirmed. *Id.* at 232, 235-38; 299 U.S. at 330-33.

6. *Curtiss-Wright* (*S.D.N.Y.*) at 240 (on rehearing).

7. *See* Schechter Poultry Corp. v. United States, 295 U.S. 495 (1935); Panama

Curtiss-Wright Export Corporation reversed the lower court decision seven to one.[8] Whatever danger the lower court decision posed to the neutrality acts was ended, but in the process several propositions were enunciated and approved that were and continue to be extremely controversial and ambiguous. Justice George Sutherland's opinion for the Court recognized sweeping federal and, more specifically, presidential power in the area of international affairs; it also seemed to free Congress from the need to formulate precise standards when it delegated power involving foreign relations.[9]

As early as 1940-1941, during the first of what Alfred H. Kelly has called the "two decisive 'breaks' in the continuity of peace-war relationships between the Executive and the Congress,"[10] *Curtiss-Wright* received attention in arguments over the nature and distribution of federal power in foreign affairs.[11] More recently, during the Vietnam War, debate over constitutional problems in warmaking has led to renewed interest in *Curtiss-Wright.* No agreement on the meaning of the case has emerged on either of these occasions. Nor has the Supreme Court come to definite conclusions about its meaning; what emerges from the cases in which the Court has cited *Curtiss-Wright* is a variety of views not unlike the range of statements from nonjudicial commentators.

Refining Co. v. Ryan, 293 U.S. 388 (1935). *See also* Carter v. Carter Coal Co., 298 U.S. 238 (1936).

8 United States v. Curtiss-Wright Export Corp., 299 U.S. 304 (1936) [*Curtiss-Wright*]. Justice McReynolds dissented without opinion and Justice Stone took no part in the case. *Id.* at 333.

9. *See id.* at 315-29.

10. Prepared Statement of Professor Alfred H. Kelly, March 9, 1971, *Hearings on War Powers Legislation Before the Senate Comm. on Foreign Relations,* 92d Cong., 1st Sess. 89 (1972) [hereinafter cited as *1971 Senate War Powers Hearings*]. Kelly identified the second break as the 1950-1951 period when President Harry S Truman, acting under his own authority as Commander in Chief and as chief foreign policymaker, committed troops to Korea and raised American force levels in Europe. *Id.* at 89-91. *Curtiss-Wright* also received some attention in this period. *See, e.g.,* HOUSE COMM. ON FOREIGN AFFAIRS, BACKGROUND INFORMATION ON THE USE OF UNITED STATES ARMED FORCES IN FOREIGN COUNTRIES, H.R. Doc. No. 127, 82d Cong., 1st Sess. 8, 11, 51 (1951). For a general discussion of the Korean War debate, see Lofgren, *Mr. Truman's War: A Debate and Its Aftermath,* 31 REV. OF POL. 223 (1969).

11. For this and the other conclusions in this paragraph see pp. 170-73 *infra.*

Altogether, on and off the bench, the decision has been alleged to provide some degree of support for a number of propositions which can be grouped as follows:

1. The United States possesses all the powers of a sovereign nation.[12] Federal authority in foreign affairs is not subject to interference from the states.[13] Even though the Constitution makes no grant to Congress, the existence of congressional power to legislate concerning foreign affairs cannot be doubted.[14] But "congressional power in this sphere . . . is limited by the First Amendment."[15]

2. The President possesses inherent power in the foreign affairs field which *Curtiss-Wright* "explicitly and authoritatively define[s]."[16] Accordingly, "the President is exclusively responsible"

12. *See* Dennis v. United States, 341 U.S. 494, 519-20 (1951) (Frankfurter, J., concurring in result); notes 13 & 14 *infra.*

13. *See* Clark v. Allen, 331 U.S. 503, 517 (1947); United States v. Belmont, 301 U.S. 324, 331 (1937).

14. *See* Perez v. Brownell, 356 U.S. 44, 57 (1958).

15. Communist Party of the United States v. Subversive Activities Control Bd., 367 U.S. 1, 96 (1961). Justice Frankfurter, speaking for the Court, made this qualification after citing *Curtiss-Wright* to support the broad proposition.

16. *See* Gravel v. United States, 408 U.S. 606, 643-44 (1972) (Douglas, J., dissenting) (maintaining, however, that *Curtiss-Wright* required such presidential power to be exercised in compliance with the Constitution); New York Times Co. v. United States, 403 U.S. 713, 728-29 & n.3 (1971) (Stewart, J., concurring); *id.* at 741-42 (Marshall, J., concurring); Youngstown Sheet & Tube v. Sawyer, 343 U.S. 579, 661 n.3 (1952) (Clark, J., concurring); Krauff v. Shaughnessy, 338 U.S. 537, 542 (1950); United States v. California, 332 U.S. 19, 45 (1947) (Frankfurter, J., dissenting); Wright, *The Transfer of Destroyers to Great Britain,* 34 AM. J. INT'L L. 680, 680-81 & n.5 (1940); 39 OP. ATT'Y GEN. 484, 486-87 (1940) (quotation in text accompanying this note is at 486); citations in notes 17-23 *infra. Cf.* Prepared Statement of Secretary of State Henry Stimson, Jan. 29, 1941, *Hearings on a Bill to Promote the Defense of the United States (S. 275) Before the Senate Comm. on Foreign Relations,* 77th Cong., 1st Sess. 90-91 (1941) (S. 275 was the Senate version of the Lend-Lease Bill, H.R. 1776); *Hearings on Lend-Lease Bill Before the House Comm. on Foreign Affairs,* 77th Cong., 1st Sess. 334 (1941) (exchange between Norman Thomas and Representative James A. Shanley, Jan. 22, 1941); Prepared Statement of Senator Barry Goldwater, April 23, 1971, *1971 Senate War Powers Hearings, supra* note 10, at 355; Acheson, *The Eclipse of the State Department,* 49 FOREIGN AFFAIRS 593, 593-94 (1971). The extent of agreement on this proposition is well illustrated by the *New York Times* case. The three justices who cited *Curtiss-Wright*—Stewart and Marshall, both of whom concurred in the Court's decision, and Harlan, who dissented—all agreed that it supported an independent executive power in foreign affairs. (Marshall thought *Curtiss-Wright* was qualified in this regard by Kent v.

for "the conduct of diplomatic and foreign affairs";[17] he is the "sole organ" of government with respect to foreign relations;[18] and he may conclude international agreements not requiring the consent of the Senate.[19] *Curtiss-Wright* raises a "serious question whether the Congress can constitutionally limit the President's powers in the field of our relations with foreign governments."[20] This interpretation also sanctions executive withholding of information pertaining to foreign affairs and national security from Congress[21] and from the public.[22] It similarly bolsters claims that the government need not reveal information related to national security, which was obtained without warrant through electronic surveillance, to defendants whose rights may be directly affected.[23]

Dulles, 357 U.S. 116 (1958), but that does not alter his assessment of the meaning of *Curtiss-Wright* itself.) *See* New York Times Co. v. United States, *supra* at 728-29, 741-42 & n.2, 756.

17. Johnson v. Eisentrager, 339 U.S. 763, 789 (1950).

18. *See* First National Bank v. Banco Nacional de Cuba, 406 U.S. 759, 766-68 (1972) (Rehnquist, J., plurality opinion); New York Times Co. v. United States, 403 U.S. 713, 756 (1971) (Harlan, J., dissenting); Hirata v. MacArthur, 338 U.S. 197, 208 (1949) (Douglas, J., concurring); Chicago & Southern Air Lines v. Waterman Corp., 333 U.S. 103, 111 (1948); United States v. Pink, 315 U.S. 203, 229-30 (1942); Memorandum entitled "Termination of Hostilities" by the Legal Advisor, Green H. Hackworth, to the Sec'y of State, Dec. 20, 1941, 1 FOREIGN RELATIONS OF THE UNITED STATES: DIPLOMATIC PAPERS 1942, at 8, 12 (1960). *Cf.* TO PROMOTE THE DEFENSE OF THE UNITED STATES, H.R. REP. NO. 18, 77th Cong., 1st Sess., pt. 1, at 6-7 (1941); Prepared Statement of Senator Barry Goldwater, *supra* note 16, at 355.

19. *See* United States v. Pink, 315 U.S. 203, 229-30 (1942).

20. Memorandum entitled "The President's Powers under H.R. 1776 [the Lend-Lease Bill] and some Prior Instances of Presidential Power," in House Comm. on Foreign Affairs, File on H.R. 1776, National Archives, Washington, D.C. This memorandum was prepared in the Treasury Department and was probably written in January 1941. I am indebted to Congressman Jerry L. Pettis of California for assistance in securing a copy of this document. *See* W. KIMBALL, THE MOST UNSORDID ACT: LEND LEASE, 1939-1941, at 174-75 & n.56 (1969).

21. *See* Prepared Statement of Dean Acheson, July 28, 1971, in *Hearings on Executive Privilege: The Withholding of Information by the Executive Before the Subcomm. on Separation of Powers of the Senate Comm. on the Judiciary*, 92d Cong., 1st Sess. 259, 260 (1971); Prepared Statement of William H. Rehnquist, Ass't Att'y Gen., July 29, 1971, *id.* at 428, 433-34.

22. *See* Brief for the United States at 13-15, New York Times Co. v. United States, 403 U.S. 713 (1971); United States v. Marchetti, 466 F.2d 1309, 1315 (4th Cir. 1972), *cert.* denied, 93 S. Ct. 553 (1973).

23. *See* Brief for the United States at 16, 31, United States v. District Court for Eastern Dist. of Michigan, Southern Division, 407 U.S. 297 (1972).

3. Legislation concerning foreign relations need not satisfy the same judicial tests applied in delegation decisions involving domestic matters.[24]

4. Those portions of Sutherland's opinion which go beyond the issue of delegation in foreign affairs are dicta.[25] The decision does not legitimate loosely controlled delegation in the domestic area even though such delegation is associated with the conduct of foreign relations;[26] it does not obviate the need for senatorial approval of agreements with foreign powers;[27] and it does not affect Congress's power to declare war.[28]

5. *Curtiss-Wright* says nothing about who is to make foreign policy (as opposed to who is to execute it).[29] In fact, *Curtiss-*

24. *See* Zemel v. Rusk, 381 U.S. 1, 17 (1965); Youngstown Sheet & Tube v. Sawyer, 343 U.S. 579, 635-36 n.2 (1952) (Jackson, J., concurring); *Ex parte* Endo, 323 U.S. 283, 298 (1944); Yakus v. United States, 321 U.S. 414, 462 (1944) (Rutledge, J., dissenting); 87 CONG. REC. 491 (1941) (Representative Bloom); *id.* at 525 (Representative Richards); *id.* at App. 280 (radio address by Senator George); R. HULL & J. NOVOGROD, LAW AND VIETNAM 173-74 (1968); Jones, *The President, Congress, and Foreign Relations,* 29 CALIF. L. REV. 565, 575 (1941); Note, *Congress, The President, and the Power to Commit Forces to Combat,* 81 HARV. L. REV. 1771, 1802 (1968); Statement by W. Rehnquist on the President's authority to order the attack on the Cambodian sanctuaries, before the Society of International Law, June 16, 1970, *reprinted in* SENATE COMM. ON FOREIGN RELATIONS, 91ST CONG., 2D SESS., DOCUMENTS RELATING TO THE WAR POWERS OF CONGRESS, THE PRESIDENT'S AUTHORITY AS COMMANDER IN CHIEF AND THE WAR IN INDOCHINA 175, 181 (Comm. Print 1970); note 25 *infra.*

25. *See* Youngstown Sheet & Tube v. Sawyer, 343 U.S. 579, 635-36 n.2 (1952) (Jackson, J., concurring); 87 CONG. REC. 1605 (1941) (Senator Wheeler); Prepared Statement by Sec'y of State William P. Rogers, May 14, 1971, *1971 Senate War Powers Hearings, supra* note 10, at 495-96 & n.47; Borchard, *The Attorney General's Opinion on the Exchange of Destroyers for Naval Bases,* 34 AM. J. INT'L L. 690, 691 (1940).

26. *See* 87 CONG. REC. 517-19 (1941) (Representative Day).

27. *See id.* at 1347-48 (Senator Shipstead); *id.* at 1604-05 (Senator Wheeler).

28. *See id.* at 1605 (Senator Wheeler); Prepared Statement of Professor Alexander Bickel, July 26, 1971, *1971 Senate War Powers Hearings* at 555. Wheeler had earlier contended that *Curtiss-Wright,* which he sometimes identified as "Wright Brothers against the United States," "contains the express language that only Congress can declare war and wage war." 87 CONG. REC. 1049 (1941). He implicitly corrected himself when he attributed (*id.* at 1605-06) the doctrine that "the term 'to declare war' necessarily connotes 'the plenary power to wage war' " to Justice Sutherland's opinion for the Court in United States v. MacIntosh, 283 U.S. 605 (1931), in which case, however, Sutherland was not carefully distinguishing between powers of the separate branches.

29. Pollak, *et al., Indochina: The Constitutional Crisis—Part 2,* 116 CONG. REC. 16478, 16480 n.13 (May 21, 1970).

Wright does not hold that looser standards are permissible in connection with delegation involving foreign affairs.[30]

On several occasions the Court has rejected broad interpretations of the foreign relations power; it has nevertheless avoided directly attacking *Curtiss-Wright*.[31] If anything, the uses to which the opinion has been put confirm that "[o]ne fact should not be overlooked–that the ambiguities of the opinion do not rule it out as available precedent."[32] In view of the Court's continued use of the case and the ongoing debate over its meaning, it is important to take a close look at what Sutherland said and at the evidence he adduced in support of his position.

I. PRECEDENT VERSUS PRINCIPLE

Curtiss-Wright held that the Joint Resolution of May 28, 1934,[33] was not an unconstitutional delegation of congressional power to the President. To understand and assess Justice Sutherland's opinion, it must be viewed historically. At the outset one must remember that the Court had recently taken a very narrow view of the permissible scope of delegatory legislation.

30. Prepared Statement of Professor Alexander Bickel, *supra* note 28, at 555; Bickel, *The Constitution and the War,* 54 COMMENTARY, July 1972, at 49, 52.

31. *See* Afroyim v. Rusk, 387 U.S. 253, 257 (1967); Kent v. Dulles, 357 U.S. 116, 129 (1958); Reid v. Covert, 354 U.S. 1, 38-39 (1957) (Black, J., plurality opinion).

32. Levitan, *The Foreign Relations Power: An Analysis of Mr. Justice Sutherland's Theory,* 55 YALE L.J. 467, 494 (1946).

33. *Resolved* . . . , That if the President finds that the prohibition of the sale of arms and munitions of war in the United States to those countries now engaged in armed conflict in the Chaco may contribute to the reestablishment of peace between those countries, and if after consultation with the governments of other American Republics and with their cooperation, as well as that of such other governments as he may deem necessary, he makes proclamation to that effect, it shall be unlawful to sell, except under such limitations and exceptions as the President prescribes, any arms or munitions of war in any place in the United States to the countries now engaged in that conflict, or to any person, company, or association acting in the interest of either country, until otherwise ordered by the President or by Congress.

> Sec. 2. Whoever sells any arms or munitions of war in violation of section 1 shall, on conviction, be punished by a fine not exceeding $10,000 or by imprisonment not exceeding two years, or both.

48 Stat. 811 (1934).

In the *Panama*[34] and *Schechter*[35] cases, the Court required that delegatory legislation specify the policy it was designed to effectuate, establish a standard to monitor subsequent executive action, and state those findings of fact the President was required to make before acting.[36] On brief in *Curtiss-Wright* the government maintained that the Joint Resolution met these tests.[37] Considering the result in the lower court and the government's and defendants' arguments on appeal, the key issue was whether the Joint Resolution required a finding of fact.[38] On this point, the government contended:

> The fact to be found by the President–whether prohibition of the sale of arms in this country may contribute to the reestablishment of peace in the Chaco–is not, as the defendants have contended, a vague matter of opinion merely, upon which only a guess might be made. On the contrary, it is an eminently practical question depending upon facts which were peculiarly available to the President . . . . The state of the war, which might vary widely and rapidly, the number and type of purchases in this country by each side, and the sales if no prohibition were imposed, which would depend upon the financial condition of the belligerents, transportation facilities, and the like, were all factors which of necessity entered into his decision. Perhaps even more important was the fact of cooperation [with other countries] . . . .[39]

These verbal gymnastics do not seem particularly persuasive in view of the *Panama-Schechter* tests. In *Schechter* the Court contended that in promulgating a Code under the National Industrial Recovery Act, one of "the finding[s] that the President is to make [is] that the code 'will tend to effectuate the policy of this title.'

34. Panama Refining Co. v. Ryan, 293 U.S. 388 (1935) [*Panama*].

35. Schechter Poultry Corp. v. United States, 295 U.S. 495 (1935) [*Schechter*].

36. *See Panama* at 414-30; *Schechter* at 530-42.

37. Brief for the United States at 10, 16-17, United States v. Curtiss-Wright Export Corp., 299 U.S. 304 (1936).

38. *See Curtiss-Wright* (*S.D.N.Y.*) at 232-36, 239; Brief for the United States at 10-17, Brief for Appellees John S. Allard, Clarence W. Webster and Samuel J. Abelow at 10-21, Brief for Appellees Curtiss-Wright Export Corp. and Curtiss Aeroplane & Motor Co., Inc. at 16-26, United States v. Curtiss-Wright Export Corp., 299 U.S. 304 (1936).

39. Brief for the United States at 11, United States v. Curtiss-Wright Export Corp., 299 U.S. 304 (1936) (citations omitted).

While this is called a finding, it is really but *a statement of an opinion* as to the general effect upon the promotion of trade or industry of a scheme of laws."[40] In consequence the NRA failed to meet one of the tests of valid delegation. The requirement in the Joint Resolution that the President find an embargo "may contribute" to the reestablishment of peace was no more precise. In addition, no hard and fast assessment could be made of those factors which the government claimed[41] would determine the effect of the embargo. A judgment based on them would have amounted to the forbidden statement of opinion.[42] Notwithstanding the view of a recent author, the Joint Resolution required the President to find more than "a necessary factual condition precedent."[43] And even if the Joint Resolution could be interpreted to require a finding, it failed to obligate the President to act once he had made the requisite finding. Contrary to the apparent requirement of *Panama* and *Schechter*[44] he retained absolute discretion to issue or not to issue a proclamation prohibiting arms shipments.[45]

So Sutherland and the Court might have accepted the government's argument that the Joint Resolution met the *Panama-Schechter* tests, but an opinion bottomed on that contention would have been shaky. Certainly a more solid foundation was available: The resolution did not have to meet the *Panama-Schechter* tests,

40. *Schechter* at 538 (emphasis added).

41. *See* p. 174 *supra*.

42. A variety of international uncertainties attended the Chaco arms embargo, which was eventually implemented to greater or lesser degree by about thirty nations. *See* W. GARNER, THE CHACO DISPUTE: A STUDY OF PRESTIGE DIPLOMACY 92-98 *passim* (1966); Hudson, *The Chaco Arms Embargo*, INT'L CONCILIATION, No. 320, at 217 (1936).

43. Bickel, *The Constitution and War*, 54 COMMENTARY, July 1972, at 49, 52. I also think Professor Bickel is incorrect in stating: "The joint resolution closely defined what the President was to do, namely stay out of war . . . ." *Id.* The Resolution said nothing about keeping the United States out of the Chaco War. Rather, the Resolution authorized the President—if he concluded it might serve the purpose of peace in the Gran Chaco—to keep American armaments out of a war which hardly threatened to engulf the United States. Nor does Professor Bickel's overall discussion of *Curtiss-Wright* support this particular statement of his. *See id.*

44. *Cf. Panama* at 430; *Schechter* at 541.

45. *See* text of the Joint Resolution: "if . . . [the President] makes proclamation . . . ." 48 Stat. 811 (1936). *See generally* Brief for Appellees Curtiss-Wright Export Corp. and Curtiss Aeroplane & Motor Co., Inc. at 16-26, United States v. Curtiss-Wright Export Corp., 299 U.S. 304 (1936).

for it fell into a category of legislation which was not governed by the tests. This category, using Sutherland's own language, consisted of legislation whose "whole aim . . . is to affect a situation entirely external to the United States, and falling within the category of foreign affairs."[46]

To establish and sanction such a category Sutherland might have followed the lead of the government on brief[47] and cited long-standing legislative and judicial precedent, but he rejected this simple course. Taking a more involved route, he began with the proposition that the discretion vested in the President was consistent with early American constitutional principles. It was in this context that Sutherland made his remarks about the inherent federal foreign relations power and the Executive's independent role in foreign affairs.[48] Only then did he review the series of "acts or joint resolutions of Congress authorizing action by the President in respect of subjects affecting foreign relations, which either leave the exercise of the power to his unrestricted judgment or provide a standard far more general than that which has always been considered requisite with regard to domestic affairs."[49] These acts,[50] he said, comprised "an impressive array of legislation . . . , enacted by nearly every Congress from the beginning of our national existence to the present day, [which] must be given unusual weight in the process of reaching a correct determination of the problem."[51] The relevant rule had been set forth in several cases.[52] As stated in *Field v. Clark,* "the practical construction of the Constitution, as given by so many acts of Congress, and embracing almost the entire period of our national existence, should not be overruled, unless upon a conviction that such legislation was clearly incompatible with the supreme law of the land."[53]

46. *Curtiss-Wright* at 315.

47. *See* Brief for the United States at 7-9, 15, United States v. Curtiss-Wright Export Corp., 299 U.S. 304 (1936).

48. *Curtiss-Wright* at 315-22.

49. *Id.* at 324.

50. *See* legislation listed in *Curtiss-Wright* at 324-27 & n.2 and additional legislation listed in *Panama* at 421-22 (which was noticed in *Curtiss-Wright* at 327).

51. *Curtiss-Wright* at 327.

52. *See* listing in *Curtiss-Wright* at 328-29.

53. Field v. Clark, 143 U.S. 649, 691 (1892).

But if legislative precedent was so compelling, why did it not dispose of the matter? Why did Sutherland not omit his sweeping remarks about the principles which informed the American Constitution in the field of foreign affairs? His biographer offers a clue, commenting that the "deductive method characterized all of Sutherland's efforts. Always there was a recurrence to first principles."[54] Such a logical style would not tolerate an opinion based on a rule derived inductively. In fact, a careful reading of the section of the opinion on legislative precedent reveals Sutherland did not think that its existence validated the Joint Resolution. His words seem chosen to convey a different meaning: Long-standing practice suggests that there exist independent constitutional justifications for the practice and these justifications are found "in the origin and [early] history of the power involved, or in its nature, or in both combined."[55]

The discussion of legislative precedent, therefore, did not so much establish a narrower ground for upholding the Joint Resolution as it simply demonstrated the working of underlying principles. The rule of interpretation which Sutherland quoted from *Field v. Clark*[56] indicated that long-standing legislative practice could be overruled if there existed "a conviction that such legislation was clearly incompatible with the supreme law of the land." Sutherland prefaced his own statement of the rule[57] with a similar comment.[58] Thus, while he later admitted that overturning the legislative practice in question was unlikely,[59] on balance he strengthened his argument—given his premises—by showing grounds inde-

54. J. Paschal, Mr. Justice Sutherland: A Man Against The State 15 (1951).

55. *Curtiss-Wright* at 328. *See, e.g.,* "The principles which justify such legislation find overwhelming support in the unbroken legislative practice which has prevailed almost from the inception of the national government to the present day," *id.* at 322; "a legislative practice such as we have here . . . goes a long way in the direction of proving the presence of an unassailable ground for the constitutionality of the practice, to be found in the origin and history of the power involved, or in its nature, or in both combined." *Id.* at 327-28. Had Sutherland meant that long-standing practice established constitutionality, he could have said this less cumbersomely.

56. *See* p. 176 *supra.*

57. *See id.*

58. *See Curtiss-Wright* at 327.

59. *See id.* at 329.

pendent of legislative precedent for upholding the Joint Resolution.

Sutherland similarly declined to ground his position on judicial precedent. The government had claimed that *Field v. Clark*[60] and *Hampton and Company v. United States*[61] "[c]learly . . . controlled" *Curtiss-Wright.*[62] "The tariff Acts there involved," said the government, "like the resolution here, were inseparably related to the external relations of the United States."[63] The government also averred that the early case of *The Aurora*[64] and other more recent cases–including *Panama*–established the propriety of executive discretion.[65] It is not difficult to infer why Sutherland omitted this argument. Judicial precedent by itself would have had the same inconclusive quality that he ascribed to legislative precedent. To paraphrase a later comment by Justice Frankfurter,[66] the Constitution and principles inherent in it, and not what the courts had ruled, remained for Sutherland "the ultimate touchstone of constitutionality."

Even if judicial precedent had deserved great weight, it did not necessarily support the government: A majority which included Sutherland had recently put its own gloss on *The Aurora, Field,* and *Hampton.* In *Panama* the Court had found that delegatory legislation had been upheld in these cases *not* because it fell into a special category of foreign relations legislation, but because the delegation involved was limited, being purely conditional in *The Aurora* and *Field* and administrative in *Hampton.*[67] The Court did hint that delegation possessed greater validity when it conferred on the President "an authority which was cognate to the conduct by him of the foreign relations of the Government."[68] The hint, however, is almost imperceptible. It was placed in a dis-

60. 143 U.S. 649 (1892).
61. 276 U.S. 394 (1928).
62. *See* Brief for the United States at 6, United States v. Curtiss-Wright Export Corp., 299 U.S. 304 (1936).
63. *Id.* at 15.
64. Brig Aurora v. United States, 11 U.S. (7 Cranch) 382 (1813).
65. *See* Brief for United States at 15-16, United States v. Curtiss-Wright Export Corp., 299 U.S. 304 (1936).
66. *See* Graves v. New York *ex rel.* O'Keefe, 306 U.S. 466, 491-92 (1939) (Frankfurter, J., concurring).
67. *See Panama* at 423-26, 429-30.
68. *Id.* at 422.

cussion of early legislative acts which, as the Court noted, "were not the subject of judicial decision"[69] and were interim or short-term in duration.[70] There were other recent cases involving delegation in foreign affairs,[71] but the government in *Curtiss-Wright* recognized that they dealt only tangentially with the constitutional issue.[72]

Finally, the opinion in *Curtiss-Wright* cannot be understood apart from Sutherland's own intellectual background. The ideas expounded in *Curtiss-Wright* about the foreign relations power, which the government only fleetingly and vaguely suggested on brief,[73] were hardly new to the Justice in 1936. In 1919 he had presented similar views in a book entitled *Constitutional Power and World Affairs*.[74] This book reiterated a thesis which he had

69. *Id.*

70. *See id.* at n.9. For further discussion of the interim and short-term nature of these early delegatory acts, see L. FISHER, PRESIDENT AND CONGRESS: POWER AND POLICY 58-61 (1972).

71. *See* United States v. Chavez, 228 U.S. 525 (1913); United States v. Mesa, 228 U.S. 533 (1913).

72. *See* Brief for United States at 15, United States v. Curtiss-Wright Export Corp., 299 U.S. 304 (1936). Interestingly, the government overlooked a pregnant dictum of Sutherland's in Carter v. Carter Coal Co., 298 U.S. 238, 295 (1936). There, in his opinion for the Court, he had argued that the federal government possessed "no *inherent* power in respect of the internal affairs of the states," but he had also observed: "The question in respect of the inherent power of that government as to the external affairs of the nation and in the field of international law is a wholly different matter which it is not necessary now to consider" (emphasis in original). Yet, while anticipating Sutherland's position in *Curtiss-Wright* later that year, this section of the *Carter* opinion was concerned with the issue of dual federalism and not with the delegation issue that the government argued in *Curtiss-Wright*. *See id.* at 289-97. The omission is thus understandable. *See also* note 81 *infra*.

73. The suggestion was made in the government's use of the quotation from *Panama* quoted at p. 178 *supra*. Brief for the United States at 8, 15, United States v. Curtiss-Wright Export Corp., 299 U.S. 304 (1936). Defendants took notice of this hint in denying that the "Constitution . . . affords any basis for the contention that the power to make laws which may affect our foreign relations may be vested in the Executive instead of in the Congress." Brief for Appellees Allard *et al.* at 10, United States v. Curtiss-Wright Export Corp., 299 U.S. 304 (1936). In district court the government had been more explicit in advancing an argument similar to the one Sutherland used in *Curtiss-Wright*. *See Curtiss-Wright* (*S.D.N.Y.*) at 239.

74. G. SUTHERLAND, CONSTITUTIONAL POWER AND WORLD AFFAIRS 25-47, 116-26 (1919) [hereinafter cited as SUTHERLAND, CONSTITUTIONAL POWER]. This book was based on Sutherland's Blumenthal Lectures at Columbia University in 1918.

advanced in an article in 1909[75] and which he possibly acquired from one of his teachers at the University of Michigan Law School.[76]

Both the book and article contended that the power of the federal government with respect to foreign relations was and always had been complete. Such power did not come from delegations by the states or from affirmative grants in the Constitution but derived from external sovereignty inherited by the federal government from Great Britain via the united colonies and the Confederation. In these writings Sutherland had been interested in refuting the doctrine that dual federalism (which he otherwise accepted) placed limitations on federal activity in the external realm. He did not devote much attention to the allocation of the general foreign relations power among the branches of the federal government.[77] He did claim, however, that *extra*-constitutional action in foreign affairs should not be equated with *un*-constitutional action.[78] In other words, without directly confronting the separation of powers issue involved in *Curtiss-Wright,* Sutherland developed the ideas he would put into service in strikingly similar language in his later opinion.[79] Thus, one source of Sutherland's comments in *Curtiss-Wright* is not difficult to discover: it was Sutherland himself. The case put the Justice in what one writer has called "the happy position of being able to give [his] writings and speeches the status of law."[80]

The fact that the views Sutherland expressed in *Curtiss-Wright* were his own nevertheless does not relegate them to status of

75. Sutherland, *Internal and External Powers of the National Government,* 191 NORTH AM. REV. 373 (1910) [hereinafter cited as Sutherland, *Internal and External Powers*]. This also appeared as S. DOC. No. 417, 61st Cong., 2d Sess. (1910).

76. *See* Paschal, *supra* note 54, at 226-28. *See also id.* at 15-20.

77. For the main exceptions to this statement, see SUTHERLAND, CONSTITUTIONAL POWER 70-91 (concerning the locus of the war power) & 122-32 (concerning participation in the treaty power).

78. *See id.* at 55; Sutherland, *Internal and External Powers, supra* note 75, at 384.

79. *See* Levitan, *The Foreign Relations Power: An Analysis of Mr. Justice Sutherland's Theory,* 55 YALE L.J. 467, 469-70, 473-76 (1946) (conveniently collecting pertinent quotations from *Curtiss-Wright* and Sutherland's book and article). Especially *compare* SUTHERLAND, CONSTITUTIONAL POWER, *supra* note 74, at 116-17, 125-26, *with Curtiss-Wright* at 316-17, 320-21.

80. Levitan, *supra* note 79, at 476.

dicta. Aside from a personal commitment to these views, he had good reasons for his opening discussion of first principles regarding the foreign relations power. Although it came first in his presentation, the discussion of principles followed from his comments about legislative precedent, and was not logically superfluous. It offered a means of establishing what resort to either legislative or judicial precedent could not conclusively establish and what respect for consistency and tenable distinctions required if the Joint Resolution were to be upheld in the face of *Panama* and *Schechter*.

II. First Principles: Justice Sutherland's Argument and Evidence

Justice Sutherland's fundamental premise was that the external and internal powers of the federal government "are different, both in respect of their origin and their nature." Internally, the government was one of enumerated powers, with the Constitution designed "to carve from the general mass of legislative powers *then possessed by the states* such portions as it was thought desirable to vest in the federal government, leaving those not included in the enumeration still in the states." However, "since the states severally never possessed international powers, such powers could not have been carved from the mass of state powers but obviously were transmitted to the United States from some other source." These "international powers" or "powers of external sovereignty" devolved on the federal government from the Confederation government, which in turn acquired them from "the colonies in their collective and corporate capacity as the United States of America." The ultimate source was the British Crown, although its possession of the external sovereignty of the American colonies ceased prior to America's formal independence because "[e]ven before the Declaration [of Independence], the colonies were a unit in foreign affairs, acting through a common agency–namely the Continental Congress . . . ." "Rulers come and go," wrote Sutherland; "governments end and forms of government change; but sovereignty survives. A political society cannot endure without a supreme will somewhere. Sovereignty is never held in suspense. When, therefore, the external sovereignty of Great Britain in respect of the colonies ceased it immediately passed to the Union."[81] To borrow a term from his earlier writings,[82] Sutherland sought to place the federal government's foreign relations power on an *extra*-constitutional footing.

After providing additional detail[83] Sutherland turned to Joseph Story's *Commentaries on the Constitution* for "general confirmation of" his position.[84] Because Sutherland saw fit to rely upon Story, the views of Story are of particular concern in understanding *Curtiss-Wright*. Story advanced three separate meanings of the term "sovereignty." "By 'sovereignty' in its largest sense," he wrote, "is meant, supreme, absolute, uncontrollable power, the *jus summi imperii,* the absolute right to govern . . . . A State which possesses this absolute power, without any dependence on any foreign power or state, is in this largest sense a sovereign state."[85] At the same time, Story noted, "the sovereignty of the government, organized within the state, may be of a very limited nature."[86] By this "sovereignty of the government," or sovereignty in "a far more limited sense," Story meant "such *political powers,* as in the actual organization of the particular state or nation are to be exclusively exercised by certain public functionaries, with-

81. *Curtiss-Wright* at 315-17. The only immediate citations for these remarks were Penhallow v. Doane, 3 U.S. (3 Dall.) 54, 80-81 (1795), which is discussed at pp. 190-91 *infra,* and Carter v. Carter Coal Co., 298 U.S. 238, 294 (1936). The latter citation is intriguing. In *Carter* Sutherland had hinted at the view he would expound in *Curtiss-Wright* (*see* note 72 *supra*), but here he used *Carter* merely to support his assertion that internally the federal government was one of enumerated powers. Sutherland's later citation of Story's *Commentaries* also covered this material.

82. *See* SUTHERLAND, CONSTITUTIONAL POWER, *supra* 74, at 55; Sutherland, *Internal and External Powers, supra* note 75, at 384.

83. *See* pp. 187-91 *infra.*

84. *Curtiss-Wright* at 317 n.1.

85. 1 J. STORY, COMMENTARIES ON THE CONSTITUTION OF THE UNITED STATES § 207, at 191-92 (1833) [hereinafter cited as STORY]. I have used the first edition of Story, which is now readily available as a reprint. The text of the sections cited by Sutherland does not vary between the first edition and the fourth edition, which Sutherland used. Pagination does vary, so I have also included section numbers to facilitate use of other editions. The sections of Story relied upon in this and the following three paragraphs are those cited by Sutherland.

86. *Id.* § 208, at 194.

out the control of any superior [governmental] authority."[87] A third sense of the term, as set forth by Story, is in "speak[ing] of a state as sovereign . . . in reference to foreign states. Whatever may be the internal organization of the government of any state, if it has the sole power of governing itself and is not dependent upon any foreign state, it is called a *sovereign state;* that is, it is a state having the same rights, privileges, and powers, as other independent states."[88]

Sutherland recognized that the united colonies, prior to independence, exercised sovereignty in its limited sense–that is, they held certain powers of sovereignty, including some pertaining to external affairs. For Sutherland, though, another *government* was evidently the only possible source for these powers of sovereignty. So, since the states severally had never possessed powers of "external sovereignty," these powers must have come from Great Britain. In turn, the schemes by which the states vested only enumerated powers in the federal government–that is, de facto arrangements until 1781, the Articles of Confederation from 1781 to 1789, and the Constitution after 1789–had little effect on the allocation and limitation of external powers.

From Story's perspective, however, there is a flaw in Sutherland's argument. For Story, the true source of the powers of sovereignty, whether external or internal, was the sovereign in the largest or absolute sense of the term. In the United States "[t]he absolute sovereignty of the nation is in the *people* of the nation . . . ."[89] The people were "the foundation, upon which the super-structure of the liberties and independence of the United States has been erected."[90] The only point at which Story came at all close to Sutherland's view of the transmission of sovereignty was in a passage he quoted from *Chisholm v. Georgia:* "From the crown of Great Britain the sovereignty of their country [that is, the United States] passed to the *people* of it . . . ."[91] Yet this es-

87. *Id.* § 207, at 192 (emphasis added).
88. *Id.* § 209, at 195 (emphasis in original).
89. *Id.* § 208, at 195 (emphasis added).
90. *Id.* § 214, at 203.
91. Chisholm v. Georgia, 2 U.S. (2 Dall.) 419, 470 (1793), *quoted* in 1 Story § 216, at 205 (emphasis is Story's and does not appear in the original).

tablishes the people as the possessors of absolute sovereignty and as the source of governmental power.[92]

Finally, the sovereignty of the United States "in reference to foreign states"—that is, in Story's third sense—also had its source in an act of the people. "The people of the united colonies," wrote Story, "made the united colonies free and independent states, and absolved them from all allegiance to the British Crown."[93] It is difficult to interpret Story as claiming that the Continental Congress held powers of "external sovereignty" by virtue of the sovereign status of the United States "in reference to foreign states." Instead, the reverse was true: The existence of powers of sovereignty, based on the consent of the absolute sovereign, gave rise to the sovereignty of the United States "in reference to foreign states." Story specifically linked Congress's authority in the external realm to the approval of people.[94]

Although Story's *Commentaries* fails to support Sutherland's position, a complete evaluation of Sutherland's account of sovereignty in early America requires consideration of the views of Americans in the 1770's and 1780's. In this regard, a recent scholar has written that James Wilson's defense, in 1785, of the constitutionality of the Bank of North America "will . . . suggest to the student of constitutional history the inherent powers doctrine of . . . *Curtiss-Wright.*"[95] Wilson claimed that Article II of the Articles of Confederation, which reserved to the states all powers not expressly delegated to the Confederation government, pertained only to those powers the states had once held. Since no state had held power to incorporate a bank commensurate with the needs of the *United* States, no state could have retained that power. Contrary to Sutherland, however, and like Story, Wilson held that the general powers of the Confederation "result[ed] from the union of the whole . . . ." He did not mention England as a possible source.[96] Moreover, the argument that certain broad pow-

92. *See also* Chisholm v. Georgia, 2 U.S. (2 Dall.) 419, 470-72 (1793).

93. 1 Story § 211, at 199.

94. *See id.* §§ 213-14, at 200-02.

95. McCloskey, *Introduction,* in The Works of James Wilson 3 (R. McCloskey ed. 1967).

96. *See* Wilson, *Considerations on the Bank of North America, in id.* at 829-30 (quotation is at 829). *See generally id.* at 824-40.

ers, not mentioned in the Articles, were vested in the Confederation government independently of state delegation would have no force concerning the foreign relations power, for the Articles gave explicit attention to foreign relations.[97] Whatever its immediate impact,[98] Wilson's defense of the bank foreshadowed the popular sovereignty position toward which he and other key figures in the 1780's were slowly moving: *Absolute sovereignty* could reside with the people, with different governments within the same nation simultaneously possessing different *powers* of sovereignty.[99]

Prior to the late 1780's few Americans accepted the popular sovereignty position, however implicit it may have been in the events and documents of independence.[100] Most of those who paid attention to political issues would have denied that from the beginning of the Revolution the people of the United States (or of the united colonies) were the constitutive body with respect to the emerging central government. They would have agreed that absolute sovereignty is indivisible, and, like William Blackstone, they would have assigned it to a particular legislative body and not to the people at large. An implication of this early position was that

97. *See* ART. OF CONFED. arts. VI & IX.

98. Professor McDonald describes Wilson as "the ablest spokesman for the theory that sovereignty devolved upon Congress or the whole people," but also claims that during the Confederation period few took Wilson's argument seriously. F. MCDONALD, E PLURIBUS UNUM: THE FORMATION OF THE AMERICAN REPUBLIC 1776-1790, at 191 n.† (1965). Certainly when the bank was chartered in 1781, both Robert Morris, the Confederation's Superintendent of Finance who proposed it, and Congress had doubts about congressional authority in the area. Congress accepted Morris's recommendation that the states be requested to pass enabling legislation. *See* Letter from Morris to the President of Congress, May 17, 1781, in 4 THE REVOLUTIONARY DIPLOMATIC CORRESPONDENCE OF THE UNITED STATES 421 (F. Warton ed. 1889); Letter from Morris to the Governors of the States, Jan. 8, 1782, in 5 *id.* at 94-95; 20 JOURNALS OF THE CONTINENTAL CONGRESS 545-48 & accompanying notes (1912) (session of May 26, 1781); 21 *id.* at 1190 (session of Dec. 31, 1781); Letter from the Virginia Delegates to the Governor of Virginia, Jan 8, 1782, in 6 LETTERS OF MEMBERS OF THE CONTINENTAL CONGRESS 288-89 (E. Burnett ed. 1921). *Cf.* Letter from Alexander Hamilton to Morris, April 30, 1781, in 2 THE PAPERS OF ALEXANDER HAMILTON 604, 630 (H. Syrett and J. Cooke eds. 1961) (tacit admission of Congress's severely limited authority). On Wilson's general relations with the Bank of North America, see C. SMITH, JAMES WILSON: FOUNDING FATHER, 1742-1798, at 140-58 (1956).

99. *See* G. WOOD, THE CREATION OF THE AMERICAN REPUBLIC 1776-1787, at 344-89, 524-47 (1969).

100. The challenge to then existing notions of legislative sovereignty is evident in, *e.g.*, the Declaration of Independence.

absolute sovereignty must rest with either the central government or the state governments, but could not rest with both.[101]

In 1776 and 1777 Congress debated the proper allocation of sovereignty;[102] the outcome is readily apparent. The second article of the draft plan of union first considered by Congress provided that each colony "reserves to itself the sole and exclusive Regulation and Government of its internal police in all matters that shall not interfere with the Articles of Confederation."[103] The plan finally approved by Congress and eventually ratified by the states contained a rather different clause: "Each state retains its sovereignty, freedom, and independence and every power, jurisdiction, and right which is not by the confederation expressly delegated to the United States, in Congress assembled."[104] This provision of the Articles, unlike the original draft, did not distinguish between powers of "internal police" and other powers. It is not surprising that in 1786 the Confederation's Secretary for Foreign Affairs, John Jay, reported to Congress that the rights to make war, peace, and treaties were a part of the "perfect though limited sovereignty" which "the thirteen independent sovereign states" had vested in Congress *"by express delegation of power."*[105] Congress subsequently approved resolutions embodying Jay's view.[106]

101. *See* G. Wood, *supra* note 99, at 344-83 *passim,* which may be usefully supplemented by two older studies: A. Small, The Beginnings of American Nationality: The Constitutional Relations between the Continental Congress and the Colonies and States (8 Johns Hopkins Studies in Historical and Political Science (H. Adams ed. 1890)); Van Tyne, *Sovereignty in the American Revolution: An Historical Study,* 12 Am. Hist. Rev. 529 (1907). *But cf.* Nettels, *The Origins of the Union and of the States,* 58 Mass. Hist. Soc. Proc. 68 (1957-60).

102. *See* M. Jensen, The Articles of Confederation: An Interpretation of the Social-Constitutional History of the American Revolution 1774-1781, at 167-76 (1959); G. Wood, *supra* note 99, at 354-61. Jensen is not particularly rigorous in distinguishing the congressional sovereignty and popular sovereignty positions, but neither were some of the people he discusses.

103. 5 Journals of the Continental Congress 547 (1906) (July 12, 1776).

104. Art. of Confed. art. II.

105. 31 Journals of the Continental Congress 797-98 (1934) (Oct. 13, 1786) (emphasis added). At issue was state compliance with the terms of the Treaty of Peace with Great Britain. Interestingly, in *The Federalist Papers,* Jay spoke of the treaty power under the Constitution as being delegated without indicating by whom. *See* The Federalist No. 64, at 432 (J. Cooke ed. 1961).

106. *See* 32 Journals of the Continental Congress 124-25, 177-84 (1936) (March 21, April 13, 1787).

As Jay's report exemplifies, contemporary comments on the issue of sovereignty were often imprecise and sometimes contradictory; formulation of a new concept of sovereignty in America was slow and unsystematic.[107] The crucial point is that both the earlier state sovereignty position and the emerging popular sovereignty position, which was arguably manifested in the Constitution,[108] are at odds with Sutherland's views.[109] Each attributes to the central government only those powers delegated by the possessors of absolute sovereignty; neither admits of extra-constitutional powers.

Sutherland nevertheless adduced some specific evidence for his position. He claimed the Treaty of 1783 between Great Britain and the "United States of America" was a "practical application of [the] fact" that external sovereignty passed immediately from Britain to the American Union.[110] However, the Articles of Confederation, which went into effect in 1781, had expressly granted to Congress the treaty-making power.[111]

The Constitution, said Sutherland, "was ordained and established . . . to form 'a more perfect Union' " in circumstances where the existing government already "was the sole possessor of external sovereignty." Such sovereignty remained "in the Union . . . without change except in so far as the Constitution in express terms qualified its exercise."[112] The implication regarding the source of the new government's external powers is as clear as it is

107. *See* notes 99 & 101 *supra,* and A. McLaughlin, A Constitutional History of the United States 131-36 (1935), which contains an especially perceptive discussion.

108. *See,* e.g., The Federalist Nos. 39-40 (J. Madison); 1 Story §§ 350-60, at 318-19; G. Wood, *supra* note 99, at 529-32, 536-47. The nature of the Constitution, of course, remained unsettled at least until the Civil War. Some still debate it. *See, e.g.,* J. Kilpatrick, The Sovereign States (1957).

109. Some commentators on *Curtiss-Wright* seem not to have grasped this, at least as evidenced by their long defenses of the state sovereignty position. *See* Levitan, *supra* note 32, at 479-89; Patterson, *In re The United States v. The Curtiss-Wright Corporation,* 22 Texas L. Rev. 286, 304-08, 445-63 (1944).

110. *Curtiss-Wright* at 317. At this point Sutherland himself placed quotation marks around the United States of America.

111. Art. of Confed. art. IX. Even so, "the peace treaty . . . was ratified by some states separately, and in New Hampshire it was ratified by individual towns." F. McDonald, *supra* note 98, at 191 n.†. For specific examples of earlier state action and claims to authority in the area of foreign relations, see Van Tyne, *supra* note 101, at 539-41.

112. *Curtiss-Wright* at 317.

misleading. In fact, the Constitution, like the Articles, mentions important aspects of the foreign relations power.[113] Thus it is more accurate to say that the power is vested in the federal government because it is explicitly and implicitly granted by the Constitution and that it ultimately derives not from Britain but from whatever body legally ordained and established the Constitution.

Sutherland's contention that the Constitutional Convention "was called and exerted its powers upon the irrefutable postulate that though the states were several their people in respect of foreign affairs were one"[114] is roughly accurate but is irrelevant to his theory. If anything, the implication that the people of the United States were the source of the foreign relations power detracts from his position. To be more accurate Sutherland might have said that the Convention was called because it was felt that the United States *should* be one with respect to foreign affairs (and certain other affairs), but that they could easily become several, and that they were most definitely having a difficult time maintaining themselves internationally as a nation.[115] This refined premise, however, also cuts against claiming an extra-constitutional origin for the foreign relations power; it indicates the Constitution was needed to establish a more effective federal power respecting foreign affairs. Indeed, in *The Chinese Exclusion Case,*[116] cited at this point by Sutherland as lending some support to his view regarding the inherent, extra-constitutional nature of the foreign relations power, Justice Field had stated: "The power of exclusion of foreigners [is] an incident of sovereignty belonging to the government of the United States, as a part of those sovereign powers *delegated* by the Constitution . . . ."[117]

113. *See* U.S. CONST. art. I, §§ 8 & 10; *id.* art. II, § 2; *id.* art. VI.

114. *Curtiss-Wright* at 317.

115. *See, e.g.,* 1 THE RECORDS OF THE FEDERAL CONVENTION OF 1787, at 19 (M. Farrand ed. 1911) (May 29, 1787) [hereinafter cited as FARRAND]; THE FEDERALIST NOS. 3 & 5 (J. Jay), 13 (A. Hamilton); P. VARG, FOREIGN POLICIES OF THE FOUNDING FATHERS 46-69 (Penguin ed. 1969); Farrand, *The Federal Convention and the Defects of the Confederation,* 2 AM. POL. SCI. REV. 532, 536 (1908); Marks, *Foreign Affairs: A Winning Issue in the Campaign for Ratification of the United States Constitution,* 86 POL. SCI. Q. 444 (1971). For a brighter view of the Confederation's prospects, see M. JENSEN, THE NEW NATION: A HISTORY OF THE UNITED STATES DURING THE CONFEDERATION 1781-1789 (1950).

116. Chae Chan Ping v. United States, 130 U.S. 581 (1889).

117. *Id.* at 609 (emphasis added). *See also* Fong Yue Ting v. United States, 149 U.S. 698, 757-58 (1893) (Field, J., dissenting).

It would seem that Sutherland found firmer support for his position in a statement he quoted from Rufus King, made during the Constitutional Convention:

> The States were not "sovereigns" in the sense contended for by some. They did not possess the peculiar features of sovereignty. They could not make war, nor peace, nor alliances, nor treaties. Considering them as political Beings, they were dumb, for they could not speak to any foreign Sovereign whatever. They were deaf, for they could not hear any propositions from such Sovereign. They had not even the organs or facilities of defence or offence, for they could not of themselves raise troops, or equip vessels, for war.[118]

An immediate problem with King's statement is a general one that arises when using the Convention's debates to interpret the Constitution. As Madison noted several years later, the state ratifying conventions and not the Philadelphia Convention gave the Constitution its legal "life and validity."[119] Moreover, whatever opinion in Philadelphia may have been, it is hardly conclusive *in itself* in determining how Americans of 1787-1788 generally interpreted the Constitution. Nor does it provide any authoritative guide to what Americans thought about the preceding decade of their history either in 1787-1788 or at the time. It may, of course, give insight into ideas and assumptions of the period,[120] which makes further examination of King's statement worthwhile.

When examined more closely, the King quotation provides scant support for Sutherland's theory, since like most of Sutherland's evidence it says nothing about whether the Confederation's power respecting foreign affairs was inherited, extra-constitutionally, from England or derived from a constitutional source.[121] Elsewhere in

118. 1 FARRAND, *supra* note 115, at 323 (June 19, 1787) (Madison's notes). Sutherland's version of the quotation, in *Curtiss-Wright* at 317, is taken from an earlier printing of Madison's notes, and differs slightly in spelling, capitalization and punctuation.

119. 5 ANNALS OF CONG. col. 776 (1849) (April 6, 1796).

120. *See* Lofgren, *War-Making Under the Constitution: The Original Understanding,* 81 YALE L.J. 672, 677-78, 690 n.79 (1972). *See also* Anderson, *The Intention of the Framers: A Note on Constitutional Interpretation,* 49 AM. POL. SCI. REV. 340 (1955).

121. "Constitutional" in this context refers, of course, not to the Constitution drafted in 1787 but to the constitutive base of the Confederation government.

the same speech, however, King clearly implied the power was delegated by the states. "If the states [under the Confederation] therefore retained some portion of their sovereignty," he commented, "they had certainly *divested* themselves of essential portions of it."[122]

Sutherland did not survey the views of other participants in the Convention. Some of them did state or imply that the states were sovereign under the Confederation.[123] Others denied this was the case,[124] but one of this group—James Madison—later in life claimed he had been misinterpreted.[125] James Wilson continued to portray the people of the United States as the source of federal authority under the Articles.[126] Most members did not discuss the issue of sovereignty. None adhered to the position advanced nearly 150 years later in *Curtiss-Wright.*

The early views of the Supreme Court should be accorded considerable weight in interpreting the Constitution and Sutherland recognized this in citing *Penhallow v. Doane.*[127] However, the opinion he cited, by Justice Patterson, was but one of four seriatim opinions in the case, and it does not support Sutherland's argument. Patterson portrayed the external power of the Continental Congress prior to the ratification of the Articles as deriving not from Great Britain or some other source external to the ordinary constitutive authority, but from the people.[128] Repeating his opin-

122. 1 FARRAND, *supra* note 115, at 324 (June 19, 1787) (Madison's notes) (emphasis added); *accord, id.* at 331 (King's notes). In his own mind King evidently thought the Confederation derived power from both the states and the people. *See* R. ERNST, RUFUS KING: AMERICAN FEDERALIST 100 (1968).

123. *See, e.g.*, 1 FARRAND, *supra* note 115, at 19, 250, 340-41.

124. *See, e.g., id.* at 324-25, 471.

125. *See* Letter from Madison to W.C. Rives, Oct. 21, 1833, in 3 *id.* at 521, 522.

126. *See* 1 *id.* at 329. See Wilson's remarks in the Pennsylvania ratifying convention, 2 THE DEBATES IN THE SEVERAL STATE CONVENTIONS ON THE ADOPTION OF THE FEDERAL CONSTITUTION . . . 432-58 *passim* (J. Elliot ed. 1888) [hereinafter cited as ELLIOT] for a more extended discussion of his views on sovereignty in late 1787.

127. 3 U.S. (3 Dall.) 54, 80-81 (1795), cited in *Curtiss-Wright* at 317. In order of Sutherland's presentation, this was the first piece of evidence cited. I have postponed discussion of it to this point in order to preserve a rough chronological framework.

128. "These high acts of sovereignty were submitted to, acquiesced in, and approved of, by the people of *America.* In Congress were vested, because by Con-

ion on Circuit, Justice Blair took a similar view, although he was ambiguous whether Congress's external authority derived directly from the people or indirectly from the people via the states.[129] Justices Iredell and Cushing did not doubt that the Continental Congress's power derived from state delegation.[130] While each Justice satisfied himself that the Continental Congress, even before ratification of the Articles of Confederation, had held sufficient authority to establish an appellate procedure in Revolutionary War Prize Cases,[131] none of them used arguments supportive of Sutherland's view.

Sutherland missed one piece of evidence from the original Constitution which could be interpreted to support his case. While ordinary legislation is the supreme law of the land if "made in Pursuance" of the Constitution, treaties become supreme law when made under the "Authority of the United States."[132] In the state ratification debates several antifederalists attacked this arrangement.[133] The federalists paid little attention to the problem, con-

gress were exercised with the approbation of the people, the rights and powers of war and peace." 3 U.S. (3 Dall.) at 80 (emphasis in original).

129. *See id.* at 109-13. For an argument basing congressional authority after independence on grounds similar to those adduced by Patterson, but introducing ambiguities not unlike those in Blair's opinion, see Ware v. Hylton, 3 U.S. (3 Dall.) 199, 222-23 (1796) (Chase, J., separate opinion).

130. *See* 3 U.S. (3 Dall.) at 92-95, 117 (separate opinions of Iredell & Cushing, JJ.). The remaining two members of the Court took no part in the decision; Chief Justice Jay was in England on the diplomatic mission which resulted in Jay's Treaty, and Justice Wilson disqualified himself because of an earlier involvement in the case. SMITH, *supra* note 98, at 372. One can guess, though, that Wilson would have written an opinion similar to Patterson's and that Jay might have done the same or else have tended toward the more ambiguous position espoused by Blair. Arguments of counsel in *Penhallow,* while coming closer to Sutherland's position than did any of the Justices' opinions, still differed from it in significant respects. *See* 3 U.S. (3 Dall.) at 74, 76.

131. Iredell took a different view of certain other aspects of the case and would have fashioned a slightly different decree.

132. U.S. CONST. art. VI.

133. *See, e.g.,* PAMPHLETS ON THE CONSTITUTION OF THE UNITED STATES 1787-1788, at 312, 331 (P. Ford ed. 1888); ESSAYS ON THE CONSTITUTION OF THE UNITED STATES 1787-1788, at 361 (P. Ford ed. 1892); 4 ELLIOT, *supra* note 126, at 215. These statements all suggest, however, that the real concern of the antifederalists was not the supremacy of treaties over the federal constitution but their supremacy over state constitutions and law—a considerably different issue. For a more direct statement indicating this concern, see, *e.g.,* 3 *id.* at 500-14. For

centrating instead on the broader issue of whether the treaty-making process, as specified in the Constitution, contained adequate safeguards for state and regional interests.[134] This fact may indicate that most federalists agreed treaties were extra-constitutional, or it may indicate that they did not regard the charge as credible enough to warrant refutation.[135] What the majority of delegates to the state conventions thought is problematic. Whether the original understanding of the status of treaties would support Sutherland is therefore also problematic.[136] By the time Sutherland wrote *Curtiss-Wright* the Court had implied that treaties were subject to the Constitution,[137] and he himself had taken the same view.[138]

---

general discussions of the treaty problem in the federal and state conventions, see W. COWLES, TREATIES AND CONSTITUTIONAL LAW: PROPERTY INTERFERENCES AND DUE PROCESS OF LAW 18-49 (1941); S. CRANDALL, TREATIES: THEIR MAKING AND ENFORCEMENT 43-63 (2d ed. 1916).

134. *See, e.g.*, ESSAYS ON THE CONSTITUTION, *supra* note 133, at 165; PAMPHLETS ON THE CONSTITUTION, *supra* note 133, at 355, 376; 2 ELLIOT 465-66, 476-77, 505-07; 3 *id.* at 292-93, 359, 516; 4 *id.* at 119-21, 280-81; THE FEDERALIST No. 75 (A. Hamilton). *See also* C. Warren, *The Mississippi River and the Treaty Clause of the Constitution,* 2 GEO. WASH. L. REV. 271 (1934).

135. It should be noted that in a few instances federalists stated or implied that treaties would be subject to the Constitution in the same way laws were. *See* 3 ELLIOT 504, 507, 514; 4 *id.* at 28, 271. It may or may not be significant that both Wilson and Madison stated during their respective state ratifying conventions that there probably would be few occasions for concluding treaties. 2 ELLIOT 513; 3 *id.* at 410.

136. Nevertheless, some indication of an original understanding (in the sense of an opinion commanding majority support) that treaties were subject to a higher authority can be gleaned from the Judiciary Act of 1789, which recognized that state courts might rule against the validity of a treaty. *See* 1 Stat. 73, 85-87 (§ 25) (1789).

137. *See* COWLES, *supra* note 133, at 292-95 *passim;* B. SCHWARTZ, A COMMENTARY ON THE CONSTITUTION OF THE UNITED STATES: THE POWERS OF GOVERNMENT 133-42 (1963). Missouri v. Holland, 252 U.S. 416 (1920), held that the Tenth Amendment was not a bar to legislation implementing a valid treaty; it thereby recognized existing authoritative opinion on the subject. *See* Boyd, *The Expanding Treaty Power,* 3 SELECTED ESSAYS ON CONSTITUTIONAL LAW 410, 422-28 (Am. Ass'n L. Schools ed. 1938). In passing, it should be noted that Professor Schwartz incorrectly concludes, at 135-36, that the wording "under the Authority of the United States" in the Supremacy Clause was included to give validity to preexisting treaties rather than to make treaties superior to the Constitution. On August 23, 1787, the Federal Convention approved the clause with wording which (among other things) provided that "all Treaties made under the authority of the U.S. shall be the supreme law . . . ." Two days later the clause "was reconsidered and after the words 'all treaties made' were inserted . . . the

In any event, based on the historical evidence outlined above, Sutherland affirmed:

> It results that the investment of the federal government with the powers of external sovereignty did not depend upon the affirmative grants of the Constitution. The powers to declare and wage war, to conclude peace, to make treaties, to maintain diplomatic relations with other sovereignties, if they had never been mentioned in the Constitution, would have vested in the federal government as necessary concomitants of nationality.[139]

He noted that the Constitution has no extra-territorial force "unless in respect of our own citizens," and thus "the operation of the nation in such territory must be governed by treaties, international understandings and compacts, and the principles of international law." Moreover, "[a]s a member of the family of nations, the right and power of the United States . . . are equal to the right and power of other members of the international family. Otherwise the United States is not completely sovereign."[140]

Here again Sutherland showed little appreciation for the several meanings of "sovereignty" which Joseph Story had discussed.[141] He also largely ignored the fact that the Constitution provides either an explicit or implicit, but still evident, authority for treaties

---

words 'or which shall be made [.]' This insertion was meant to obviate all doubt concerning the force of treaties preexisting, by making the words 'all treaties made' to refer to them, as the words inserted would refer to future treaties." 2 FARRAND, *supra* note 115, at 389, 417 (Madison's notes). What was therefore crucial to clarifying the meaning of the clause with respect to the continued validity of preexisting treaties was the insertion of "or which shall be made" and not the use of "under the Authority of the United States," which appeared in both the unclear early version and in the final version. It may well be that Schwartz (who follows Justice Black in Reid v. Covert, 354 U.S. 1 16-17 (1957), on this point) is correct that the Supremacy Clause was not intended to give treaties an extra-constitutional status, but the evidence he adduces does not prove it. Professor Corwin fell into the same trap. *See* E. CORWIN, THE PRESIDENT: OFFICE AND POWERS 1787-1957, at 421 n.17 (1957).

138. *See* SUTHERLAND, CONSTITUTIONAL POWER, *supra* note 74, at 141-65 *passim*. In fact, Sutherland anticipated Holmes's argument in Missouri v. Holland, 252 U.S. 416 (1920), but with additional emphasis that such an argument is not tantamount to putting treaties above the Constitution. *See* SUTHERLAND, CONSTITUTIONAL POWER 153-58.

139. *Curtiss-Wright* at 318.

140. *Id.*

141. *See* pp. 182-84 *supra*.

and other international understandings and compacts.[142] A similar problem arises in his interpretation of the cases[143] in which, he claimed, the Supreme Court had recognized that the United States Government held specific powers relating to foreign relations by virtue of its sovereign status under international law. These cases do state that the American nation possesses the powers incident to any sovereign nation. They also suggest that those powers reside in the federal government (or the President) by virtue of constitutional grants.[144]

Sutherland concluded his discussion about the nature of federal power in foreign affairs with a quotation from the then-recent case of *Burnet v. Brooks.* "As a nation with all the attributes of sovereignty," Chief Justice Hughes had argued, "the United States is vested with all the powers of government necessary to maintain an effective control of international relations."[145] This case, considered alone, came closest to supporting Sutherland's thesis about an inherent federal foreign relations power. Hughes recognized only such limitations to federal authority in foreign affairs as were imposed by the Constitution.[146] But *Burnet* included no extended discussion of the point, comparable to that found in *Fong Yue Ting;*[147] and as authority for the statement quoted by Sutherland,[148] Hughes relied on *Fong Yue Ting*[149] and a concurring

142. U.S. CONST. art. I, §§ 8 & 10; *id.* art. II, § 2.

143. Altman & Co. v. United States, 224 U.S. 583, 600-01 (1912); Fong Yue Ting v. United States, 149 U.S. 698, 705 *passim* (1893); Jones v. United States, 137 U.S. 202, 212 (1890); all cited in *Curtiss-Wright* at 318.

144. *See* Altman & Co. v. United States, 224 U.S. 583, 600-01 (1912); Fong Yue Ting v. United States, 149 U.S. 698, 704-12 (1893); Jones v. United States, 137 U.S. 202, 223 (1890). Among these, *Fong Yue Ting* contained the most extensive discussion of the issue; here the Court, per Justice Gray, reviewed in detail the specific constitutional provisions from which the federal foreign relations power derives. *See* 149 U.S. at 711-12. Contrary to Sutherland's implication, *Altman,* strictly speaking, did not involve the constitutional problem of the extent of the federal government's authority to conclude international agreements other than treaties, but rather the problem of interpreting the meaning of "treaties" within § 5 of the Circuit Court of Appeals Act of 1891, 26 Stat. 826, 828. *See* 224 U.S. at 600-01.

145. Burnet v. Brooks, 288 U.S. 378, 396 (1933), *quoted in Curtiss-Wright* at 318.

146. 288 U.S. at 400.

147. Fong Yue Ting v. United States, 149 U.S. 698, 711-15 (1893).

148. *See* text accompanying note 145 *supra.*

149. *See* 288 U.S. at 396.

opinion in *The Legal Tender Cases.*[150] Justice Gray's opinion for the court in *Fong Yue Ting* had traced the federal foreign relations power to specific constitutional grants.[151] Justice Bradley's concurrence in *The Legal Tender Cases* had based the federal powers over war and foreign relations on specific grants and on the fact that the states were forbidden to enter the field. Bradley spoke of "inherent" powers, but he seems to have been referring to such powers as were regarded as necessary to a general government when the Constitution was adopted and thus were implicitly vested in the federal government by the Constitution.[152]

The federal foreign relations power now tottering on an extra-constitutional footing, Sutherland asserted that "participation in the exercise of the power is significantly limited."[153] In its exercise the President was the key figure, having important roles in treaty-making and foreign negotiation generally.[154] Sutherland spoke of "the very delicate, plenary and exclusive power of the President as the sole organ of the federal government in the field of international relations–a power which does not require as a basis for its exercise an act of Congress, but which, of course, like every other governmental power, must be exercised in subordination to the applicable provisions of the Constitution."[155] Sutherland supported this view with several pieces of historical evidence.

One item was John Marshall's statement in the House of Representatives that "[t]he President is the sole organ of the nation in its external relations, and its sole representative with foreign nations."[156] Since these words of Marshall have often been quoted, it is worthwhile to put them in context. At issue was whether Pres-

150. Knox v. Lee and Parker v. Davis, 79 U.S. (12 Wall.) 457, 554 (1871) (Bradley, J., concurring).

151. *See* 149 U.S. at 711-12.

152. *See* 79 U.S. (12 Wall.) at 556. In a similar fashion, the Court in *The Legal Tender Cases* traced Congress's authority to make Treasury notes legal tender (1) to powers enumerated in the Constitution, (2) to powers implied by those enumerated powers, and (3) to *"powers [which] were understood by the people who adopted the Constitution to have been created by it,* [although such] powers [were] not enumerated and not included incidently in any of those which were enumerated . . . ." *Id.* at 534 (emphasis added).

153. *Curtiss-Wright* at 319.

154. *See id.* at 319-21.

155. *Id.* at 320.

156. 10 Annals of Cong. col. 613 (1851) (March 6, 1800), *quoted in Curtiss-Wright* at 319.

ident John Adams had acted properly in extraditing a British subject to England on a murder charge pursuant to the Jay Treaty of 1795. After the statement just quoted, Marshall continued:

> Of consequence, the demand of a foreign nation can only be made on [the President].
>
> He possesses the whole Executive power. He holds and directs the force of the nation. Of consequence, any act to be performed by the force of the nation is to be performed through him.
>
> He is charged to execute the laws. A treaty is declared to be law. He must then execute a treaty, where he, and he alone, possesses the means of executing it.
>
> The treaty, which is a law, enjoins the performance of a particular object. The person who is to perform this object is marked out by the Constitution, since the person is named who conducts the foreign intercourse, and is to take care that the laws be faithfully executed. The means by which it is to be performed, the force of the nation, are in the hands of this person. Ought not this person to perform the object, although the particular mode of using the means has not been described? *Congress, unquestionably, may prescribe the mode, and Congress may devolve on others the whole execution of the contract;* but, till this be done, it seems the duty of the Executive department to execute the contract by any means it possesses.[157]

Marshall, in other words, was claiming that the President's power could range from ministerial to discretionary, depending on what Congress had or had not done. It is difficult to extract from Marshall's comments an endorsement of unlimited executive discretion in foreign *policy*-making.

Another purported piece of evidence was an 1816 report of the Senate Foreign Relations Committee on instructing the President concerning a commercial treaty with Great Britain.[158] Stressing that the President was "the constitutional representative with regard to foreign nations," the Committee held that Senate interference threatened the design, secrecy and dispatch necessary for successful negotiations and hence would impair national security.[159] Taken as a whole, however, the Report included a mixture

157. 10 ANNALS OF CONG. cols 613-14 (emphasis added).
158. *See Curtiss-Wright* at 319.
159. Foreign Relations Comm. Report of Feb. 15, 1816, in 8 COMPILATION OF

of grounds for not instructing the President. For one thing, the President already knew the Senate's sentiments on the subject.[160] For another, the proposed Senate resolutions only duplicated past diplomatic instructions.[161] Further, the Report justified the Senate's noninvolvement in the negotiation process by noting "that if any benefits be derived from the division of the legislature into two bodies, the more separate and distinct in practice the negotiating and treaty ratifying [*sic*] powers are kept, the more safe the national interests."[162] The Committee thereby praised the requirement of *independent* Senate approval. Surely this indicated a conclusion that *if* the President had any independent authority, it extended only to the process of negotiation. Even the portion of the Report quoted by Sutherland affirmed that the President, in his conduct of foreign relations, "is responsible to the Constitution."[163] The Committee implicitly asserted that, pursuant to the Constitution, the Senate could instruct the President: instruction was simply unwise on prudential grounds.[164] (On other occasions, in fact, the Senate did advise on treaties, independently of consenting to them.)[165]

In 1796 the House of Representatives requested that it be given documents relating to the Jay Treaty before it appropriated funds for implementing the Treaty. Sutherland approvingly quoted[166] President Washington's denial of the request, a denial which echoed the claim in *Federalist* No. 64 that sharing information with the House might compromise the secrecy requisite to successful negotiations.[167] But again, this "evidence" on balance detracts

---

REPORTS OF THE COMM. ON FOREIGN RELATIONS, UNITED STATES SENATE 1789-1901, S. DOC. No. 231, 56th Cong., 2d Sess. 28 (1901).

160. *See id.*

161. *See id.* at 23.

162. *Id.* at 24-25.

163. *Id.* at 24.

164. *See id.* at 23-25.

165. *See* CRANDALL, *supra* note 133, at 72-74; Q. WRIGHT, THE CONTROL OF AMERICAN FOREIGN RELATIONS 248 (1922) (containing several instances not mentioned by Crandall). Sutherland had previously recognized a senatorial role in initiating treaties. *See* SUTHERLAND, CONSTITUTIONAL POWER, *supra* note 74, at 123.

166. *See Curtiss-Wright* at 320-21.

167. *See* Message to the House of Representatives, March 30, 1796, 1 MESSAGES AND PAPERS OF THE PRESIDENTS 186-87 (J. Richardson comp. 1897); THE FEDERALIST No. 63 (J. Jay).

from the Justice's position. Washington's reply turns out as a claim not for independent presidential authority, but for the independence of the treaty-making power–that is, the President *and* the Senate. Washington not only did not view his foreign relations power as resting on an extra-constitutional base, but he specifically linked it to the Constitution and the intentions of its framers and adopters.[168] When Sutherland commented that "the wisdom of [Washington's refusal to provide information to the House] was recognized by the House itself and has never since been doubted,"[169] he was wrong. The House debate which followed Washington's response showed the President had not convinced some members. Resolutions reaffirming the House's position passed by a 57-to-35 margin.[170] A half century later Representative John Quincy Adams conceded that the memory of Washington was "reverenced next to worship" but still argued that "the President was wrong in that particular instance [in 1796] and went too far to deny the power of the House . . . ."[171] During the century following the 1796 episode, on several occasions the House or individual committees or representatives asserted that body's prerogatives with respect to treaties and foreign affairs.[172]

The congressional practice of *requesting* the State Department to furnish information "if not incompatible with the public interest," while *directing* other Departments to do so, claimed Sutherland, evidenced recognition of "[t]he marked difference between foreign affairs and domestic affairs . . . ."[173] The uncertainties attending foreign situations, plus the need sometimes to condition

168. *See* 1 MESSAGES AND PAPERS OF THE PRESIDENTS, *supra* note 167, at 187-88. Washington's view that the state ratifying conventions agreed that the House would have no independent judgment regarding treaties is questionable, although no definitive assessment of the original understanding on this point is possible. Wilson and Madison in their state conventions pictured the House as having an "influence" on treaty-making. *See* 2 ELLIOT, *supra* note 126, at 507; 3 *id.* at 347. *But cf.* THE FEDERALIST No. 75, at 506-07 (A. Hamilton). Other federalists suggested that making commercial treaties would require the participation of the House. *See* 3 ELLIOT 365; 4 *id.* at 48, 267.

169. *Curtiss-Wright* at 320.

170. *See* 5 ANNALS OF CONG. cols. 762-83 (1849) (March 31-April 7, 1796).

171. CONG. GLOBE, 29th Cong., 2d Sess. 167 (Jan. 13, 1848).

172. *See* 2 A. HIND, PRECEDENTS OF THE HOUSE OF REPRESENTATIVES OF THE UNITED STATES, H.R. DOC. No. 355, 59th Cong., 2d Sess. 979-88, 1006-27 (1907).

173. *Curtiss-Wright* at 321.

congressionally-authorized presidential action in foreign affairs on confidential information, further indicated "the unwisdom of requiring Congress in this field of governmental power to lay down narrowly definite standards by which the President is to be governed."[174] These remarks, however, took Sutherland from an exposition of the original theory of the Constitution–that is, from first principles, which might admittedly be disclosed by *early* practice–to *general* practice and considerations of prudence. Yet, if general practice controlled the issue, his entire exposition of first principles was unnecessary. If prudence argued for imposing looser standards on foreign affairs delegation, then the issue was delegation and not independent presidential authority. Sutherland's overall point seems precisely to have been that it was the extra-constitutional nature of federal power in foreign affairs which allowed prudential considerations to override normal limitations on such power. The ultimate question must be whether the Justice had already succeeded in demonstrating the extra-constitutional nature of federal power in foreign affairs.

Undaunted, Sutherland closed his exposition of constitutional historical premises by quoting from *Mackenzie v. Hare:* "As a government, the United States is invested with all the attributes of sovereignty. As it has the character of nationality it has the powers of nationality, especially those which concern its relations and intercourse with other countries. *We should hesitate long before limiting or embarrassing such powers.*"[175] In *Mackenzie,* though, the issue of which branch might exercise certain powers was missing as was the delegation issue. In addition, the quoted section of *Mackenzie* rested on simple assertion and tended, as did Sutherland himself,[176] to confuse the powers of sovereignty possessed by a government, sovereignty in what Justice Story had called its absolute sense, and sovereignty as a relationship between nations. More important, the section was dicta. Immediately after the words quoted by Sutherland, Justice McKenna had continued: "But [such] monition is not necessary in the present case. There

174. *Id.* at 321-22.

175. Mackenzie v. Hare, 239 U.S. 299, 311 (1915), *quoted in Curtiss-Wright* at 322 (emphasis added by Sutherland).

176. *See* pp. 182-84, 193-94 *supra.*

need be no dissent from the cases cited by the plaintiff [who had been stripped of her citizenship after she married a foreigner]; there need be no assertion of very extensive power over the right of citizenship or of the imperative imposition of conditions upon it."[177] Mrs. Mackenzie had argued that only voluntary expatriation can divest one of citizenship. McKenna found that her voluntary act of marrying a foreigner in full knowledge of existing law constituted voluntary expatriation.[178] *Mackenzie* thus gave minimal support to Sutherland's position.

III. THE VERDICT ON *Curtiss-Wright*

Attempting to assess *Curtiss-Wright's* impact is hazardous. A recent writer holds that the tradition of "foreign policy . . . [being] set aside as a Presidential preserve . . . owes much to the *Curtiss-Wright* decision . . . ."[179] In discussing Franklin D. Roosevelt's actions prior to American entry into World War II, Arthur Schlesinger, Jr. contends that while avoiding "grandiose claims of executive authority," F.D.R. was "doubtless encouraged by Justice Sutherland and the *Curtiss-Wright* decision . . . ."[180] Nevertheless, it is not clear that the decision greatly changed the nation's direction. The apparent demise of the *Panama-Schechter* doctrine as regards domestic legislation[181] indicates there is little reason to conclude that an adverse decision would necessarily or likely have had a lasting impact on the permissible limits of dele-

177. Mackenzie v. Hare, 239 U.S. 299, 311 (1915).

178. *See id.* at 310-12.

179. L. FISHER, PRESIDENT AND CONGRESS: POWER AND POLICY 207 (1972).

180. Schlesinger, *Congress and the Making of American Foreign Policy,* 51 FOREIGN AFFAIRS 78, 92 (1972). "That there was not a more vigorous and sustained challenge to the constitutionality of the lend-lease measure during the course of its enactment [in 1941] can be explained only by reference to the scope of the delegation of power in the field of foreign relations sanctioned by [*Curtiss-Wright*]." Jones, *supra* note 24, at 574.

181. *See* Jaffe, *An Essay on Delegation of Legislative Power,* in SELECTED ESSAYS ON CONSTITUTIONAL LAW 1938-1962, at 89, 116-19 (Comm. of Ass'n of Amer. Law Schools ed. 1963); L. FISHER, *supra* note 179, at 71-74. In large part because of Professor Jaffe's comments, I have used the wording "apparent demise."

gation in foreign affairs. It probably would not even have affected the neutrality legislation of the latter 1930's.[182] More generally, extensive presidential power can be upheld without reference to *Curtiss-Wright*.[183] And the existence of *Curtiss-Wright* has not kept Congress from contributing to the formulation of foreign policy–a fact which Senator J. William Fulbright recognized as early as 1961.[184]

There remains another sort of historical judgment to render on *Curtiss-Wright*. The decision has had, and may continue to have, importance within particular legal and political controversies. By being available as authoritative precedent, it decreases the need to confront directly certain basic constitutional issues. As described at the beginning of the article,[185] the decision has in fact been used to support several propositions pertaining to the Constitution

182. The main "fact" which the neutrality acts required the President to find was the existence of a state of war among foreign nations. This is arguably more specific than the "fact" that an American embargo might contribute to reestablishment of peace. Moreover, once the President found that war existed abroad, these Acts (except for the 1935 Act) gave him little discretion concerning the actions he was required to take. *See* Act of November 4, 1939, 54 Stat. 4 (1939); Act of May 1, 1937, 50 Stat. 121 (1937); Act of February 29, 1936, 49 Stat. 1152 (1936); Act of August 31, 1935, 49 Stat. 1081 (1935). In practice, of course, President Roosevelt retained a measure of discretion—for example, by simply failing to "find" that China and Japan were at war. *See* R. DIVINE, THE ILLUSION OF NEUTRALITY 200-19 (1962).

183. *See e.g.*, Prepared Statement and Testimony of Professor John Norton Moore, April 25, 1972, in *Hearings on Congressional Oversight of Executive Agreements Before the Subcomm. on Separation of Powers of the Senate Comm. on the Judiciary*, 92d Cong., 2d Sess. 140, 149-60, 173-74 (1972); Prepared Statements of John R. Stevenson, Legal Advisor, Dep't of State, and William H. Rehnquist, Ass't Att'y Gen., July 1, 1970, in *Hearings on Congress, The President, and the War Powers Before the Subcomm. on National Security Policy and Scientific Developments, House Comm. on Foreign Affairs*, 91st Cong., 2d Sess. 205-16 (1970). But arguments which run counter to *Curtiss-Wright* may more generally call into question other sweeping assertions of executive prerogative in foreign affairs.

184. *See* Fulbright, *American Foreign Policy in the 20th Century under an 18th Century Constitution*, 47 CORNELL L.Q. 1, 3-6 (1961). In 1961 Senator Fulbright, on balance, still lamented the congressional input. For a discussion of Congress's impact on foreign policymaking in recent years, see FISHER, *supra* note 179, at 212-35; F. WILCOX, CONGRESS, THE EXECUTIVE AND FOREIGN POLICY (1971).

185. *See* pp. 170-73 *supra*.

and foreign affairs. Whether it actually supports these propositions can now be assessed.[186]

That the United States possesses all the powers of a sovereign nation is undoubtedly correct. That the federal government thereby *inherently* holds these powers or holds them at all does not automatically follow. Far from supporting the contention that external sovereignty devolved on the federal government ultimately from Great Britain and hence has an extra-constitutional base, Sutherland's historical evidence and judicial precedents suggest the opposite: Federal power in foreign affairs rests on explicit and implicit constitutional grants and derives from the ordinary constitutive authority. Whether Americans in 1787-1788 and earlier regarded this authority as resting with the states or with the people of the United States is not entirely clear, for there was no universally accepted view of the matter. Neither alternative provides an extra-constitutional base for the foreign relations power. One need not rely on *Curtiss-Wright* to limit state participation in foreign affairs: the Constitution imposes severe limitations.[187] Congressional authority finds similar explicit and implicit bases in the Constitution. Sutherland himself admitted that federal power in foreign relations was limited by specific prohibitions in the Constitution.

Sutherland uncovered no constitutional ground for upholding a broad, inherent, and independent presidential power in foreign relations. So, since an extra-constitutional base for the general foreign affairs power is also missing, no basis exists for concluding that the Constitution's allocation of foreign relations power between the branches may be constitutionally ignored. This is not to say that a clear allocation emerges from the Constitution or that any concrete understanding respecting allocation in the field of foreign affairs existed in 1787-1788. However, from both Suther-

186. The following assessments rest on the detailed analysis of *Curtiss-Wright* already presented, so except where additional evidence is introduced I have not deemed it necessary to include further documentation.

187. *Cf.* E. Corwin, The President: Office and Powers 1787-1957, at 173 (1957). In the same discussion, however, Corwin mistakenly equates Sutherland's theory in *Curtiss-Wright* with Justice Patterson's in Penhallow v. Doane. *See* pp. 190-91 *supra.*

land's evidence and from other sources, hints of an implicit understanding do emerge. Americans of that day probably accorded Congress a coordinate, if not a dominant, role in the initiation of war, whether declared or not.[188] Control of commercial policy was largely assigned to Congress,[189] and contemporaries thought that commercial relations would constitute a major portion of America's overall relations with the world.[190] Treaty-making involved both the President and the Senate; some Americans in 1787-1788 may have thought that the House would also have an input into treaty-making. John Marshall, at least in 1800, evidently did not believe that because the President was the sole organ of communication and negotiation with other nations, he became the sole foreign policy-maker. Marshall indicated that Congress could modify the President's diplomatic role. Sutherland made no direct claim in *Curtiss-Wright* that the President possessed authority to conclude international agreements other than treaties. He simply stated that authority existed for the United States as a nation to do so—a contention that his evidence neither proved nor disproved.[191] Sutherland said little that bears on the question of executive privilege vis-à-vis Congress. Certainly the evidence he reviewed gives no support to claims that the Executive has an inherent power to maintain the confidentiality of information in its possession.

Sutherland adduced no evidence, other than practice, that restrictions imposed by the Constitution on delegation of legislative power do not apply equally to delegation involving both domestic

188. *See* Lofgren, *supra* note 120. *But cf.* THE FEDERALIST No. 70, at 471 (A. Hamilton): "Energy in the executive is a leading character in the definition of good government. It is essential to the protection of the community against foreign attacks . . . ." However, besides making a claim which is quite narrow on its face, Hamilton in context is *not* defending the President against Congress. His purpose, instead, is to defend the executive office as established by the Constitution against antifederalist arguments for a plural Executive. *See id. passim.*

189. *See* U.S. CONST. art. I, § 8.

190. *See, e.g.,* F. GILBERT, THE BEGINNINGS OF AMERICAN FOREIGN POLICY: TO THE FAREWELL ADDRESS *passim* (1965); VARG, *supra* note 115, at 1-69.

191. For sounder arguments, see W. MCCLURE, INTERNATIONAL EXECUTIVE AGREEMENTS (1941); McDougal & Lans, *Treaties and Congressional-Executive or Presidential Agreements: Interchangeable Instruments of National Policy,* in M. MCDOUGAL, H. LASSWELL, W. BURKE, F. FELICIANO, L. LIPSON, STUDIES IN WORLD PUBLIC ORDER 404 (1960).

and foreign affairs. His evidence, other than practice,[192] leads to precisely the opposite conclusion.

In view of the doctrinal climate of the mid-1930's respecting delegation and Sutherland's comments on the need to go beyond practice to constitutional principles, there is no basis for regarding as dictum *Curtiss-Wright*'s contention that federal power involving foreign affairs rests on a different base than federal power in domestic affairs. Similarly, its contention about independent presidential power is not dictum. If practice does not conclusively establish the Constitution's meaning, advancing either or both of these contentions was not superfluous to upholding the validity of the Chaco arms embargo. Quite the contrary: These contentions are necessary elements in Sutherland's opinion.[193] On its face, moreover, *Curtiss-Wright* arguably loosens restrictions on domestic delegation when such delegation is necessary to the conduct of foreign relations.[194] It contains grounds which obviate the need for senatorial approval of certain international agreements. It does not narrow Congress's power to declare war, for Sutherland recognized that specific constitutional provisions placed restrictions on the exercise of external powers, but otherwise it implicitly supports executive authority to use the armed forces in implementing foreign policy objectives.

If one not only accepts Sutherland's premise about the need to resort to first principles rather than practice in constitutional interpretation, but also tests the historical accuracy of Sutherland's evidence, *Curtiss-Wright* does not support the existence of an extra-constitutional base for federal authority, broad independent executive authority, or laxness in standards governing delegation.

192. For a brief discussion of the support found in actual practice for Sutherland's view of presidential power, see Prepared Statement of Professor Alfred H. Kelly, April 25, 1972, in *Hearings on Congressional Oversight of Executive Agreements, supra* note 183, at 176, 178.

193. These contentions would not be dicta even if practice alone did establish Sutherland's conclusion. "[W]here a decision rests on two or more grounds, none can be relegated to the category of *obiter dictum.*" Woods v. Interstate Realty Co., 337 U.S. 535, 537 (1949) (citations omitted). *See generally* Note, *Dictum Revisited,* 4 STAN. L. REV. 509 (1952).

194. The resolution and presidential actions upheld in *Curtiss-Wright* placed restrictions on domestic activities, that is, on arms sales in the United States to foreign countries, in an effort to influence a foreign situation.

It certainly invests the President with no sweeping and independent *policy* role. If, conversely, one casts aside the notion that first principles are controlling, then *Curtiss-Wright's* comments about actual practice provide support for the proposition that delegation in foreign affairs need not be so strictly controlled as Sutherland himself probably thought domestic delegation should be. This, though, is not to say that the past practice reviewed in *Curtiss-Wright* or the legislation therein at issue involved delegation broader than the domestic delegation which has been upheld since 1936.

In sum, it is incorrect to dismiss major segments of *Curtiss-Wright* as dicta. But the history on which those segments rest is "shockingly inaccurate."[195] If good history is a requisite to good constitutional law, then *Curtiss-Wright* ought to be relegated to history.

195. I borrow this phrase from Murphy, *Time to Reclaim: The Current Challenge of American Constitutional History,* 69 AM. HIST. REV. 64, 76 (1963).

6

MR. TRUMAN'S WAR:
A Debate and Its Aftermath

"We doubt very much if there is any question in the minds of the majority of the people of this country that the conflict now raging in Korea can be anything but war," wrote Federal District Judge Harry C. Westover in the spring of 1953. "Certainly those who have been called upon to suffer injury and maiming, or to sacrifice their lives," he continued, "would be unanimous in their opinion that this is war—war in all of its horrible aspects."[1] What common sense dictated, the judge did not deny. Yet, although federal and state courts in this and several other instances touched tangentially on the question of whether the Korean conflict was a war in a legal and technical sense, on balance they produced no definitive answer. The United States Supreme Court, moreover, refused to hear any of the cases in question.[2] But the lower courts of the nation were not alone in their diversity of opinion regarding the status of the war. The political branches of the Federal Government similarly displayed little agreement on the parallel issue of what the legal basis was for American participation in the Korean venture.

But the policy-makers who ran the war and the Congressmen who voted supplies for it debated a number of issues relating to

1. *Weissman* v. *Metropolitan Life Ins. Co.,* 112 F. Supp. 420, 425 (S.D. Cal. 1953).

2. A. Kenneth Pye, "The Legal Status of the Korean Hostilities," *The Georgetown Law Journal,* XLV (Fall, 1956), 45-47; the relevant cases are cited in *ibid.,* notes 3-16.

the legal status of the conflict. These included: did United States intervention constitutionally rest on the authority of the United Nations? Or, did it rest on the President's powers as Commander-in-Chief? Or was the intervention essentially illegal in the absence of a congressional declaration of war? The participants in the debate arrived at no firm conclusions, but whatever their lack of consensus they spoke and acted as if they thought that the question of the domestic legal basis of the war was an important one.[3] Moreover, in the years since 1950, the debate and memories of it would seem to have had some impact on the handling of several other important problems in American foreign policy.

II

The major events of the first week of the Korean War are reasonably familiar. After the North Korean attack on South Korea on June 24, 1950 (Washington time), the United States quickly brought the matter to the attention of the United Nations Security Council. In the absence of the Soviet delegate, the Council speedily adopted an initial resolution which termed the attack "a breach of the peace" and called on "the authorities of North Korea to withdraw their armed forces to the thirty-eighth parallel." In addition, the Council asked "all Members to render every assistance to the United Nations in the execution of this resolution and to refrain from giving assistance to the North Korean authorities."[4] The invading forces ignored the Security Council's plea, and in Washington on June 27, 1950, President Harry S Truman ordered American air and sea forces to go to the defense of the Republic of Korea.

In announcing his action, Truman noted that the Security

3. While the author is aware of no systematic study of the 1950-1953 debate over the issues presented in the text, several specialized studies and texts do briefly consider the constitutional status of the war, either by pointing to its legal ambiguity or by stating (or implying), with varying degrees of clarity, that the American intervention legally rested on the resolutions of the United Nations Security Council or on the authority of the President as Commander-in-Chief or on both.

4. U.S. Congress, House of Representatives, *Background Information on Korea* [Documents], H. Rept. 2495, 81st Cong., 2d sess., 1950, pp. 44-45.

Council had "called upon all members of the United Nations to render every assistance to the United Nations in the execution of [the June 25th] resolution." He then stated: *"In these circumstances* I have ordered United States air and sea forces to give the Korean Government troops cover and support." Thereby he committed American forces to Korea, and, as the italicized phrase indicates, he did so in a manner suggesting that at least part of his authority for the move came from the recommendation of the Security Council. And later on June 27, the United Nations seemed to figure even more in the developing course of events. In the face of reports that the North Koreans were not withdrawing in response to the resolution of June 25, the Security Council formally recommended "that the Members of the United Nations furnish such assistance to the Republic of Korea as may be necessary to repel the armed attack and to restore international peace and security in the area." Moreover, on June 28, Secretary of State Dean Acheson told a news conference that after the first meeting of the Security Council on June 25, "all action in Korea [had] been under the aegis of the United Nations," and added: "That is a very important point."[5]

In the face of these developments, it is little wonder that there was some initial confusion within Congress regarding the constitutional basis for the President's ordering American troops into Korea. On Monday, June 26, for example, Senators William F. Knowland (R., Calif.) and Tom Connally (D., Tex.) both demonstrated their implicit beliefs that the United Nations somehow was providing the basis for the developing American response, although neither one explained precisely what the connection was. Several Congressmen expressed similar sentiments on the morning of June 27 when Truman privately informed them of his plan to send American sea and air forces to Korea.[6]

5. *Public Papers of the Presidents of the United States* (Washington; Government Printing Office, 1958-), *Harry S. Truman . . . 1950,* p. 492 (italics added) (hereafter cited as *Public Papers* with appropriate President and year); *Background Information on Korea,* p. 48; *United States Department of State Bulletin,* XXIII (July 3, 1950), 6.

6. *Congressional Record,* 81st Congress, 2d session, pp. 9154-60; Harry S. Truman, *Memoirs,* II: *Years of Trial and Hope* (New York, 1965), 384-85; Beverly Smith, "The White House Story: Why We Went to War in Korea,"

Later on the 27th, when Senator Scott Lucas (D., Ill.) read to the Senate the President's public statement announcing that he was committing United States forces, a brief but revealing debate followed. James P. Kem (R., Mo.) asked if Truman had "arrogated to himself the authority of declaring war." Lucas assured him "that history will show that on more than 100 occasions in the life of the Republic the President as Commander-in-Chief has ordered the fleet or troops to do certain things which involved the risk of war." Then, in reponse to questions from Senators Arthur V. Watkins (R., Utah) and John W. Bricker (R., Ohio), Lucas introduced the almost inevitable ambiguity into the discussion when he said that the President's move was based on the cease-fire order of the Security Council. The two Republicans seemed to agree, albeit begrudgingly, that somehow Truman derived his authority from the Council's resolution. A bit later Watkins disapprovingly observed: "Now, according to the action taken, by the mere order of the United Nations our troops can be sent into a fighting war without Congress saying 'yes' or 'no.' " Of the several Senators who participated in the exchange, only Wayne Morse (R., Ore.) dwelt explicitly on the importance of the President's powers as Commander-in-Chief. These, Morse suggested, were the powers Truman was drawing on, for they were "very broad powers in times of emergency and national crisis."[7]

Morse was on the right track, for on the following day, June 28, Senator Robert A. Taft (R., Ohio) outlined some serious objections to the notion that Truman gained constitutional authority to act by virtue of the Security Council's resolutions. It was obvious, the Ohioan thought, that the President had "brought about a *de facto* war with the government of northern Korea . . . without consulting Congress and without congressional approval." And given this fact, he contended, the whole situation took on increased urgency, because if the President's action were allowed to stand, he could proceed to commit American forces anywhere in the world. "If the incident is permitted to go by without pro-

---

Saturday Evening Post, CCXXIV (Nov. 10, 1951), 82; Glenn Paige, *The Korean Decision, June 25-30, 1950* (New York, 1968), pp. 187-91, and *passim* for other early views on the relation of the U.N. to the Korean operation.

7. *Cong. Rec.* 81st Cong., 2d sess., pp. 9228-43.

test, at least from this body," Taft warned, "we could have finally terminated for all time the right of Congress to declare war, which is granted to Congress alone by the Constitution of the United States."[8]

Taft recognized that in the past Presidents had "at times intervened with American forces to protect American lives or interests." But he did not discuss in detail what precedent these incidents provided; neither did he discuss in theoretical terms the powers of the President as Commander-in-Chief. He simply denied that, even as Commander-in-Chief, a President had "any right to precipitate any open warfare" and, in his opinion, this denial ruled out the developing intervention in Korea.[9] This aspect of Taft's argument should be kept in mind, for around it at a later time serious debate centered. In short, it was a debatable point and, in addition, Taft sketched the main elements of an almost irrefutable argument against the major alternative legal ground for Truman's action.

The major alternative was the notion that Truman gained legal authority by virtue of the Security Council's actions. In fact, Taft placed his major emphasis on the problem of what authority the Chief Executive possessed as a result of the membership of the United States in the United Nations. He conceded that the President could use American forces at the request of the Security Council and in the absence of a congressional declaration of war, but he could do this only after certain conditions had been met. In the present situation, the most relevant of these conditions was the fact that such forces could be used only if they had been previously made available to the Council in accordance with special agreements as provided in the Charter. These agreements, furthermore, required congressional approval "by appropriate act or joint resolution," according to the United Nations Participation Act of 1945. No such agreements had been negotiated with the Security Council, let alone approved by Congress. The President thus drew no power to act from the resolutions of the Security Council.[10]

Taft did not elaborate on his argument, but is was a sound one.

8. *Ibid.*, pp. 9322-23.
9. *Ibid.*
10. *Ibid.*, p. 9323.

Quite clearly the resolutions of the Security Council provided no substitute for a declaration of war in terms of domestic constitutional law. The Senate, of course, had ratified the Charter of the United Nations as a treaty, and, as such, the Charter was a part of the supreme law of the land, according to most interpretations. But the only power possessed by the Security Council to order the armed forces of U.N. members into action came from Article 42 of the Charter, which specified that the Council might use certain national forces for peace-keeping purposes. These forces, in turn, were to be furnished under the provisions of Article 43, to which Taft had alluded. Article 43, however, stated:

> All Members of the United Nations, in order to contribute to the maintenance of international peace and security, undertake to make available to the Security Council, on its call and in accordance with a special agreement or agreements, armed forces, assistance, and facilities, including rights of passage, necessary for the purpose of maintaining international peace and security. . . . The agreement or agreements shall be negotiated as soon as possible on the initiative of the Security Council. They shall be concluded between the Security Council and Members or between the Security Council and groups of Members and *shall be subject to ratification by the signatory states in accordance with their respective constitutional processes.* [Italics added.]

Moreover, as Taft had contended, the United Nations Participation Act of 1945, which had been passed to implement the non-self-executing provisions of the Charter within the United States, had to be considered. Section 6 of the Act provided that no agreements under Article 43 could become operative without approval of Congress. Interestingly enough, Dean Acheson, who at the time was Under-Secretary of State, had mentioned this feature of the Act when he testified in late 1945 before the Foreign Affairs Committee of the House of Representatives. American troops, he assured the Congressmen, would not be assigned to the U.N. for peace-keeping operations without congressional approval. Later, in 1949, Assistant Secretary of State Dean Rusk reiterated the same point during hearings on amending the Act.[11] In any case,

11. United Nations Participation Act, 59 Stat. 619, 621; Acheson's testimony is quoted in *Cong. Rec.*, 82d Cong., 1st sess., p. 5079; Rusk's is in U.S. Congress,

Taft was correct in saying that at the time of the Korean attack the Security Council had not concluded any agreements under Article 43.

In his speech of June 28, 1950, Taft did not cite all the evidence just presented. But as time passed, others restated Taft's case, especially after the dramatic reversal in the fortunes of war between September and December, 1950. Some added little detail, but two Senators, especially, provided the documentation for the case Taft had sketched against the idea that the President derived authority from the United Nations to commit American forces to Korea. In May, 1951, Senator Karl E. Mundt (R., S.D.) argued that the United States had complete freedom to press for total victory in Korea without reference to the wishes of the United Nations, and, in the process, he fully analyzed the implications of Article 43 and of the United Nations Participation Act. Later in the year in an article in *The Western Political Quarterly,* Senator Watkins reiterated the arguments of Taft and Mundt, and concluded that in fact the United States was involved in "war by executive order."[12]

No one convincingly challenged the arguments of Taft, Mundt, and Watkins regarding the United Nations. That fact, however, did not prevent a good many loose references to the conflict in Korea as being a United Nations war. President Truman typically characterized it as such, and during the first week of the war when a reporter asked him if it would be correct to call the conflict "a police action under the United Nations," he replied, "Yes. That is exactly what it amounts to." In the following months some critics

---

House, *To Amend the United Nations Participation Act of 1945,* Hearings before the Committee on Foreign Affairs, 81st Cong., 1st sess., 1949, pp. 51, 75-78. On agreements under Art. 43 see also U.S. Congress, Senate, *The Charter of the United Nations,* Hearings before the Committee on Foreign Relations, 79th Cong., 1st sess., 1945, revised edition, pp. 290-300. The 1949 amendment to the United Nations Participation Act provided that, upon the request of the U.N., the President could provide up to 1000 American military personnel at any one time to the international organization "to serve as observers, guards, or in any other *non-combatant* capacity" (italics added). 63 Stat. 734, 735-36.

12. *Cong. Rec.,* 82d Cong., 1st sess., pp. 5078-83; Arthur V. Watkins, "War by Executive Order," *The Western Political Quarterly,* IV (December, 1951), 539-49. Taft's most explicit statement of his position on the legality of the war was in his book *A Foreign Policy for Americans* (Garden City, 1951), pp. 21-36.

came to use "police action" almost as a term of derision, but even after the Eisenhower victory in November, 1952, Truman defended the term, contending the war still was a "[p]olice action for the United Nations–to stop aggression–and nothing else." If others in and out of the Administration did not use "police action" with quite so much abandon, they, too, referred to the war as somehow being a United Nations operation. In opposition to Taft's view, to cite one instance, *The Nation* quoted Harold Stassen, the perennial G.O.P. presidential candidate, as saying that "Mr. Truman's action is in accordance with the United Nations Charter."[13]

III

Yet, if the President's critics demonstrated the weakness of the position that Truman directly derived authority to commit American forces to Korea from the Security Council's resolutions of June 25 and 27, 1950, the conclusion that Truman's authority was completely unrelated to the actions of the Council did not follow. This was made clear in a lengthy memorandum issued by the State Department ten days after the North Korean invasion.[14]

The memorandum quoted Acheson's statement, of June 28, that after June 25 "all action in Korea has been under the aegis of the United Nations," but it ultimately defended Truman's action as being a proper exercise of his powers as Commander-in-Chief.

> The President, as Commander in Chief of the Armed Forces of the United States, has full control over the use thereof. He also has authority to conduct the foreign relations of the United States. Since the beginning of United States history, he has upon numerous occasions utilized these powers in sending armed forces abroad.

To bolster this claim of presidential power, the memorandum cited statements by a variety of men from both public life and the aca-

13. Truman, *Public Papers, 1950,* pp. 504-05; *ibid., 1952-53,* p. 1058; *The Nation,* CLXXII (January 13, 1951), 21.

14. "Memorandum of July 3, 1950, Prepared by the Department of State on the Authority of the President To Repel the Attack in Korea," in *Background Information on Korea,* pp. 61-68.

demic world, and listed in an appendix 85 instances of the President's use of American armed forces without a declaration of war.

The document further contended that because the North Korean attack was in violation of the United Nations Charter and in defiance of the Security Council's resolutions, it constituted a threat to the continued existence of the international organization "as a serious instrumentality for the maintenance of international peace." And this, in turn, affected the United States, because the preservation of the United Nations as a force for peace in the world was a "cardinal interest of the United States." In addition, the attack endangered another American interest in that it posed a direct threat to the security of American forces in the Pacific area. The memorandum concluded:

> These interests of the United States are interests which the President as Commander in Chief can protect by employment of the Armed Forces of the United States without a declaration of war. It was they which the President's order of June 27 did protect. This order was within his authority as Commander in Chief.

Thus the State Department's position, which at first glance seemed at odds with Truman's view of the situation, was actually consistent with it. The State Department did not claim that the United Nations was an irrelevant factor in assessing the domestic legality of the American involvement in Korea, and Truman, in his statements about Korea being a United Nations war, had not said that the Security Council's resolutions directly provided a legal basis for his commitment of forces to Korea. There was, in short, a difference between acting *"in support of* the authority of the United Nations," as Acheson put it on one occasion,[15] and acting *under* the authority of the world body within the context of domestic constitutional law. Further, to the extent that the position stated in the memorandum of July 3, was the considered position of the Administration, the critics of conducting the war without a declaration were beating a straw man in demonstrating how Article 43 of the United Nations Charter had not been implemented. The Administration did not claim it as a source of power.

Significantly, if Dean Acheson's testimony during the Senate

15. *Dept. of State Bulletin,* XXIII (July 10, 1950), (italics added).

hearings which followed the dismissal of General MacArthur is any indication, the position set forth in the memorandum of July 3 was the considered position of the Administration. Two occasions during his testimony were especially revealing. One involved the attempt of Senator Guy Gillette (D., Iowa) to get the Secretary of State to admit that the President was "not relying on his traditional authority as Commander in Chief of the United States forces," but rather was "acting under the authority of the United Nations Charter." Acheson persistently refused to accept Gillette's argument, and instead continued to stress that Truman was acting under his authority as Commander-in-Chief to protect the interests of the United States–interests which included the preservation of international law as represented by the U.N. In the course of countering Gillette's position, the Secretary explained that when he had said "all action in Korea has been under the aegis of the United Nations," he essentially meant that the United Nations had simply requested its members to act.[16]

The other occasion involved an attempt by Senator Harry F. Byrd (D., Va.) to force Acheson to concede that the President should have sought a declaration of war in connection with the Korean intervention. Acheson admitted to Byrd that "in the ordinary popular sense" the conflict in Korea was a war, whereupon the Senator observed: "Now, the Constitution provides, of course, as you know, that Congress is the only agency of the Government that can declare war." But Acheson steadfastly maintained that Truman had "ample" powers and precedents to cover what he had done in Korea without asking Congress for a declaration of war. As evidence, he cited both the appendix of the memorandum of July 3 and another list of about 150 past military actions.[17]

As early as July 3, 1950, then, the crux of the debate was the

16. U.S. Congress, Senate, *Military Situation in the Far East,* Hearings before the Committees on Armed Services and Foreign Relations, 82d Cong., 1st sess., 1951, pp. 1936-41.

17. *Ibid.,* pp. 2013-17. See also pp. 1767-69, 1818-19, 2018-21, and 2282-86 for additional comments by Acheson on Truman's authority to commit American forces to Korea. The list of over 150 military actions which Acheson mentioned is in a very competent study done by the staff of the Committee on Foreign Affairs of the House of Representatives: *Background Information on the Use of United States Armed Forces in Foreign Countries,* H. Rept. 127, 82d Cong., 1st sess., 1951, pp. 55-62.

issue of how extensive were the President's powers as Commander-in-Chief. On July 5, Senator Paul H. Douglas (D., Ill.) claimed that they were at least broad enough to cover the situation developing in Korea. By this time, it should be noted, the President had committed American ground troops as well as sea and air forces. Douglas could not agree with what he thought was the basic contention of Truman's critics on the legality question: ". . . that for us to use [any] armed force against persons or forces of other countries is in effect a declaration of war and hence should only be authorized by act of Congress." He noted that the Constitutional Convention had substituted the word "declare" for "make" in the clause giving Congress the power "to declare war," and he held that this was done because "the convention did not want to tie our country's hand by requiring congressional assent for all employment of armed force." He approvingly quoted James Madison's view that the change in wording left "to the Executive the power to repel sudden attacks."[18]

Douglas also defended a broad interpretation of the President's powers as being consonant with the need for rapid action. In 1950, he admitted, Congress could assemble more rapidly than previously, but war itself had become swifter. Debate, especially in the Senate, could prove lengthy in a day when "even the slightest delay may prove fatal." Finally, a broad interpretation was consistent with precedent. Douglas' list of past examples of Executive action, which was typical of the lists cited by others, included the undeclared naval war with France from 1798 to 1800, the expedition against the Barbary pirates in 1804, Polk's occupation of the area between the Nueces River and the Rio Grande, the expedition to China during the Boxer Rebellion, several Latin American incidents, and the 1919 expedition to Siberia. He conceded that in a few of these instances "the exercise of this power was probably unwise," but in most "it was distinctly wise, benefiting both this country and the world as a whole." In Korea, of course, the primary threat had not been to American lives and property, as it "almost invariably" had been in the cases cited. But Douglas said that the attack there clearly represented a threat to American

18. *Cong. Rec.*, 81st Cong., 2d sess., pp. 9647-48.

security, because it could have led to a chain of events reminiscent of those of the late 1930's.[19]

Previously, Senator Knowland had also indicated that the urgency of the situation in Korea added to the President's powers as Commander-in-Chief, but five days after Douglas spoke, Senator Alexander Wiley (R., Wisc.), the ranking Republican on the Foreign Relations Committee in the absence of ailing Arthur Vandenberg (R., Mich.), still concurred in Taft's suggestion that the proper course for the Chief Executive would be to seek formal congressional approval of his actions, even if it meant approval after the fact.[20] The President did not do this, and after MacArthur's success at Inchon in mid-September, criticism of the Administration's handling of the war diminished. With the reverses of late November and December, however, Truman again came under attack.

In January, 1951, Taft and Douglas restated their earlier positions on the legality of the conflict in Korea, and others added their comments. But it remained for Senator Watkins, on January 22, to dissect thoroughly the precedents which Douglas had cited in July.

> Careful examination and study will bring out the fact [he said] that practically without exception each one of the incidents cited in the various lists is clearly distinguishable from, and not in point with, the elements of fact and law surrounding the Korean intervention.

He also charged that Douglas had misinterpreted the intentions of the framers of the Constitution in substituting "declare" for "make" in the clause giving Congress the power "to declare war." As Watkins saw it, the framers "proposed to empower the Executive to take action necessary to the immediate self-defense of the Nation in the event of sudden attack. It was not their intention, however, to empower the Executive to declare war; nor was it

19. *Ibid.*, pp. 9648-49. Douglas also described the legal significance of the resolutions of the Security Council. He contended that although they provided Truman with no direct authority in domestic constitutional terms, they guaranteed that Truman's action had the sanction of international law.

20. *Ibid.*, pp. 9320, 9538-40, 9737.

their intention to empower the Executive to make war except in self-defense of the Nation . . . ." The Korean conflict, in Watkins' view, was something different from a war "in self-defense of the Nation."[21]

IV

Senator Taft undoubtedly reflected the fears of many of his countrymen when he wrote that if the actions of President Truman in Korea were allowed to go unchallenged, then the President had "almost unlimited power to do anything in the world in the use of either troops or money." Certainly Dean Acheson was not very convincing in replying "I can't answer that . . . I don't know" when Senator Byrd asked him, in June, 1951: "Have we ever had [another] war in our history for this length of time without a declaration of war, where we have had casualties of 170,000 [*sic*]?" Yet, as Acheson, Senators Douglas and Morse, and others pointed out, there were pragmatic reasons for not declaring war. A show of force without a declaration might cause the enemy to back down, which he could not easily do if he were confronted with a formal declaration of war. Moreover, in the case of Communist China, one had to consider where a declared war would lead and how one declared war against an unrecognized power.[22]

So a concern for the future affected the protagonists' reading of history. But their very disagreement over the legality of the war, combined with the fact that they took their positions seriously, in turn had an impact on several significant episodes in American foreign policy during the period of 1950 to 1953 and in the ensuing years.

For one thing, the "Great Debate" of late 1950 and early 1951 over sending additional troops to Europe was probably a good deal more bitter because of Truman's failure to get formal congressional approval for the Korean intervention. Although one issue in the debate was whether the United States should send the additional forces, another was whether the President had the au-

21. *Ibid.,* 82d Cong., 1st sess., pp. 513-19.

22. Taft, *A Foreign Policy for Americans,* p. 33; *Military Situation in the Far East,* pp. 2017, 2282-86; *Cong. Rec.,* 81st Cong., 2d sess., p. 9648; *ibid.,* 82d Cong., 1st sess., p. 257.

thority to send them without the approval of Congress. On January 5, 1951, Taft warned:

> The President has no power to agree to send American troops to fight in Europe in a war between the members of the Atlantic Pact and Soviet Russia. Without authority he involved us in the Korean War. Without authority he apparently is now attempting to adopt a similar policy in Europe. This matter must be debated by the people of this country if we are to maintain any of our constitutional freedoms.

But in a news conference a week later, Truman himself emphatically denied that he needed the permission of Congress before sending new forces abroad. In the meantime Senator Kenneth S. Wherry (R., Nebr.) had introduced a resolution stating that it was "the sense of the Senate that no ground forces of the United States should be assigned to duty in the European area for the purposes of the North Atlantic Treaty pending the formation of a policy with respect thereto by the Congress."[23]

Even a cursory reading of the *Congressional Record* demonstrates that Taft and Wherry had considerable company in their views. The Wherry resolution itself went to the Foreign Relations and Armed Services committees, which held hearings on it through most of February, 1951. But the hearings, too, revealed a good deal of bitterness obviously linked with Korea. At one point Knowland caustically remarked, with clear reference to the meeting on June 27, 1950, between Truman and several congressional leaders just before the Korean intervention was publicly announced:

> I think some of us who have been here for a time in Washington have seen both the ranking Members of the majority and minority leadership called to the White House presumably for consultation when, as a matter of fact, it has been nothing more than showing them a press release an hour before the release is issued.

The effect of Korea could also be seen on the Administration side of the Great Debate. Secretary of State Acheson provided the

23. *Ibid.*, pp. 59, 94; Truman, *Public Papers, 1951,* pp. 19-22.

committees with a substantial brief on the powers of the President to use American armed forces abroad. On examination, interestingly, the brief turns out to follow word for word the State Department's memorandum of July 3, 1950, on the power of the President to commit forces to Korea, with only the specific references to the Korean situation omitted.[24]

The committees finally reported a watered down resolution approving the sending of four more divisions to Europe, but their resolution was by no means a blank check for the President. It stated the sense of the Senate to be not only that the President should consult with the relevant committees of both houses of Congress before committing troops to NATO, but that no ground troops in addition to the four divisions should be sent to Europe "without further Congressional approval." The resolution passed, 69 to 21, with practically all of the negative votes cast by members who thought it too weak a check on the President.[25] The profound disagreements over the legality of the Korean intervention were beginning to cast a shadow.

In a sense, too, the lesson to be on guard against Executive usurpation, which some Congressmen derived from the Korean intervention, was an underlying factor in the thinking of those who supported the Bricker Amendment. Judging by their comments, admittedly, the adherents of the proposed constitutional amendment, which would have given Congress greater control over the President's treaty-making power, saw it more as a means of guaranteeing that no future President would repeat the alleged mistakes made at Teheran, Yalta, and Potsdam. But when Senator Bricker himself introduced his amendment in its 1952 version, while the Korean War still raged, he put it this way: *"With every day bringing evidence that the issues of foreign policy are matters of life and death,* it is more important than ever to make the determination of foreign policy, insofar as practicable, the responsibility of the President and all 531 Members of Congress."[26]

24. *Cong. Rec.*, 82d Cong., 1st sess., pp. 55-3283 *passim;* U.S. Congress, Senate, *Assignment of Ground Forces of the United States to Duty in the European Area,* Hearings before the Committees on Foreign Relations and Armed Services, 82d Cong., 1st sess., 1951, pp. 88-93, 594.

25. *Cong. Rec.* 82d Cong., 1st sess., pp. 3282-83.

26. *Ibid.*, 2d sess., p. 912 (italics added).

And the same lesson was visible in the 1952 Republican Party platform. "We charge," said the Republicans, "that [the Democrats] have plunged us into war in Korea without the consent of our citizens through their authorized representatives in the Congress . . . ."[27]

Hostilities in Korea ended in July, 1953, but memories of the war apparently lingered on. On January 24, 1955, President Dwight D. Eisenhower requested Congress to confirm his authority to use American forces to block any Chinese Communist invasion of Formosa or the Pescadores Islands. While not denying that, as Commander-in-Chief, he had the authority anyway, Eisenhower held that congressional action would make the American position less ambiguous and would thus reduce the chance of misjudgment on the part of the Chinese Communists. Several days of debate followed in Congress, most of it focusing on the international ramifications of the requested resolution. But Senator Margaret Chase Smith (R., Me.) pointed to another factor. After contending that "the first and basic point of honesty to ourselves and our people" was to acknowledge "the inescapable fact" that the call for congressional action arose at least in part from domestic political considerations, Mrs. Smith said that Eisenhower's request was conditioned by all the criticism that his predecessor had sustained for conducting an undeclared war. "This resolution, when passed," she said, "will foreclose against any such similar criticism of President Eisenhower, because he will have put Democrats, as well as Republicans, in the same position with him."[28]

Senator H. Alexander Smith (R., N.J.) agreed with Mrs. Smith, and columnist Arthur Krock suggested that Eisenhower was redeeming the Republican platform's implied pledge against further undeclared wars. Sherman Adams, Eisenhower's "assistant president," later confirmed that the Korean experience was a factor in Eisenhower's decision. Certainly, the decision to ask Congress for what critics called a blank check was consistent with the views of Secretary of State John Foster Dulles, who some time earlier had

27. Kirk H. Porter and Donald B. Johnson, comps., *National Party Platforms, 1840-1956* (Urbana, 1956), p. 497.

28. Eisenhower, *Public Papers, 1955,* pp. 207-11; *Cong. Rec.,* 84th Cong., 1st sess., p. 764.

been asked whether he thought Truman should have sought a declaration of war during the Korean fighting. On that occasion, Dulles said that the wisdom of Truman's move was a difficult thing to judge after the fact, but he also testified that "so far as the future is concerned, if there should be a recurrence of the circumstances which arose in Korea, I would personally be disposed to advise the President to seek action by the Congress."[29]

Two years later, in January, 1957, Eisenhower again asked Congress to reaffirm his authority to deploy American armed forces abroad—this time in the Middle East. As in 1955, several Congressmen dwelt on the Korean experience, both during the floor debate and in committee. Moreover, Secretary Dulles emphasized to the Senate Foreign Relations Committee that, whatever the precedents for executive action without congressional approval, "President Eisenhower is very reluctant to use the Armed Forces of the United States in a way which could engage the United States in a war unless he has the authority of the Congress." "And I think, frankly," he rather pointedly added, "that the President . . . is more scrupulous, holds that view more strongly, perhaps, than some other Presidents have done." It may also be of some significance that, during its hearings on the Middle East Resolution, the House Foreign Affairs Committee showed a special interest in former Secretary of State Acheson's reflections on the constitutional aspects of the Korean situation.[30]

The next major application of the lesson of the Korean debate came in the autumn of 1962. With reports publicly circulating

29. *Ibid.*, p. 823; *The New York Times,* January 25, 1955; Sherman Adams, *Firsthand Report: The Story of the Eisenhower Administration* (New York, 1961), p. 130; U.S. Congress, Senate, *Statements of Secretary of State John Foster Dulles and Adm. Arthur Radford, Chairman, Joint Chiefs of Staff,* Hearings before the Committee on Foreign Relations, 83rd Cong., 2d sess., 1954, pp. 37-38; Formosa Resolution, 69 Stat. 7.

30. Eisenhower, *Public Papers, 1957,* pp. 6-16; *Cong. Rec.,* 85th Cong., 1st sess., pp. 1186, 1201, 1207, 2400, 2531-32, 2946-47, 3025-26; U.S. Congress, Senate, *The President's Proposal on the Middle East,* Hearings before the Committees on Foreign Relations and Armed Services, 85th Cong., 1st sess., 1957, pp. 273, 307, 308-12, 379-80, 888-89, 903-04; U.S. Congress, House, *Economic and Military Cooperation with Nations in the General Area of the Middle East,* Hearings before the Committee on Foreign Affairs, 85th Cong., 1st sess., 1957, pp. 189-90, 196-97; Middle East Resolution, 71 Stat. 5.

that the Soviets were building Cuba into an offensive base, and after several statements on the subject by President John F. Kennedy, Congress formally resolved that the United States was determined not to allow Cuba to become a threat to hemispheric security. This time, in fact, the initiative for the resolution came not from the President, but from Congress. Kennedy had even declared that, as Commander-in-Chief, he already possessed the authority to check the Cuban buildup, using force if necessary. During the Senate hearings on the resolution, however, several Senators referred back to the Korean crisis and implied that President Truman would have been wise to have obtained congressional approval for the American intervention. Had that been done, Richard B. Russell (D., Ga.) maintained, "we wouldn't have had a political issue that has been going around ever since then." Consistent with Russell's view, the committee report on the resolution stated that such a resolution made "it possible to avoid constitutional arguments over relative powers of the President and the Congress respecting the use of American Armed Forces."[31]

It remained for Lyndon B. Johnson to make the clearest admission that he thought the Korean debate demonstrated the political necessity for obtaining congressional approval of troop commitments abroad. In early August, 1964, after a North Vietnamese attack on American naval units in the Gulf of Tonkin, Johnson asked for and obtained a resolution stating that "the Congress approves and supports the determination of the President, as Commander-in-Chief, to take all necessary measures to repel any armed attack against the forces of the United States and to prevent further aggression." At the time, Johnson explicitly linked the resolution to the need for national unity with far greater emphasis than had Eisenhower when he requested the Formosa or Middle East resolution. Concurrently, Secretary of State Dean Rusk testified that the Formosa, Middle East, and Cuba resolu-

31. Kennedy, *Public Papers, 1962,* pp. 674-75; U.S. Congress, Senate, *Situation in Cuba,* Hearings before the Committees on Foreign Relations and Armed Services, 87th Cong., 2d sess., 1962, pp. 35-36, 59-60, 77-78; U.S. Congress, Senate, *Situation in Cuba,* S. Rept. 2111, 87th Cong., 2d sess., 1962, p. 2; Cuba Resolution, 76 Stat. 697.

tions "form[ed] solid legal precedent for the action now proposed."[32]

By mid-1967, of course, Johnson's increasing commitment of forces to Vietnam was coming under heavy attack, especially in the Senate. In this atmosphere, during a news conference on August 18, 1967, the President developed with considerable frankness the connection between his request, three years before, for the Southeast Asia resolution and the Korean debate. After dwelling on Senator Taft's objection to the lack of formal congressional participation in the Korean decision, Johnson detailed how, following the Tonkin Gulf attack, he had consulted with various congressional leaders. As he recalled the episode: "We reviewed with them Senator Taft's statements about Korea, and the action that President Eisenhower had taken, and asked their judgment about [a] resolution that would give us the opinion of Congress." Out of the consultation, said Johnson, came the Southeast Asia resolution. On August 21, 1967, three days after the President's news conference, Under-Secretary of State Nicholas Katzenbach summarized Johnson's remarks in this way for the Foreign Relations Committee: "As the President said on Friday, it was in an effort to eliminate [a Korean-type] debate that the Tonkin Gulf resolution was proposed and passed and supported."[33]

V

Because the Truman Administration did not seriously maintain that the various resolutions of the Security Council provided a substitute for a congressional declaration of war, the issue in the debate over the legality of the Korean intervention was whether the President had properly exercised his powers as Commander-in-Chief. It would be too much to claim that the desire to avoid

32. Johnson, *Public Papers, 1963-64,* pp. 930-32, 938, 940, 946; U.S. Congress, Senate, *Southeast Asia Resolution,* Hearings before the Committees on Foreign Relations and Armed Services, 88th Cong., 2d sess., 1964, p. 3; Southeast Asia Resolution, 78 Stat. 384.

33. U.S. Congress, Senate, *U.S. Commitments to Foreign Powers,* Hearings before the Committee on Foreign Relations, 90th Cong., 1st sess., 1967, pp. 125-26 (news conference transcript), 130-31.

similar debates was the only factor underlying the blank-check resolutions of 1955, 1957, 1962, and 1964. Yet, each of these resolutions provided an occasion for Congress or the President or both to reflect on the legality of the Korean venture. As the years pass and more archival and manuscript materials become available, it will be of some interest to determine more precisely just how great was the influence of memories of the Korean experience in comparison with, say, the influence of the international situation at the time each resolution was passed.

Some commentators, of course, have suggested that President Truman would have been well advised to seek formal congressional approval of *his* intervention and, by thereby tying the potential critics in with Administration policy, eliminate future criticism. Since 1953, only in one case have sustained combat operations followed a blank-check resolution. In the period since the adoption of the Southeast Asia resolution, the United States has become increasingly involved in Vietnam. But at the same time criticism of the war has steadily mounted. From this, one might conclude in retrospect that Truman would not have guaranteed himself continuing support simply by securing congressional approval in late June or early July of 1950, when he probably could have easily obtained it. In other words, the actual controversy stemmed more from the implications of the highly ambiguous nature of modern limited war rather than from Truman's failure to obtain congressional assent. Both sides in the Korean debate conceded that the President could act, without Congress, to counter an immediate, dangerous threat to American interests and security. Thus the real issue became (and remains): What constitutes such a threat? To answer that question takes one beyond the province of constitutional law.

Index